CIVIL BECOMINGS

NGOgraphies: Ethnographic Reflections on NGOs

The NGOgraphies book series explores the roles, identities, and social representations of nongovernmental organizations (NGOs) through ethnographic monographs and edited volumes. The series offers detailed accounts of NGO practices, challenges the normative assumptions of existing research, and critically interrogates the ideological frameworks that underpin the policy worlds where NGOs operate.

CIVIL BECOMINGS

Performative Politics in the Amazon
and the Mediterranean

RAÚL ACOSTA

The University of Alabama Press   Tuscaloosa

The University of Alabama Press
Tuscaloosa, Alabama 35487-0380
uapress.ua.edu

Typeface: Minion Pro

Cover images: Mediterranean Social Forum panel, Barcelona (*top*), and workshop
for farmers on controlling fires, Brazil (*bottom*); photos courtesy of Raúl Acosta
Cover design: Michele Myatt Quinn

Cataloging-in-Publication data is available from the Library of Congress.
ISBN: 978-0-8173-2067-6
E-ISBN: 978-0-8173-9319-9

Contents

Figures

Acknowledgments

The research that informs this book was made possible by two generous grants: the Philip Bagby Studentship, from the Institute of Social and Cultural Anthropology of the University of Oxford, in the United Kingdom, and the Researchers Training Fund, from the Technical and Higher Studies Western Institute (ITESO), in Guadalajara, Mexico.

Short stays at Humboldt University, in Berlin, funded by the German Academic Exchange Service (DAAD) and the Mexican Association of Universities (ANUIES), at the University of Aberdeen, funded by the University of Deusto, and at the Max Planck Institute for the Study of Religious and Ethnic Diversity, in Göttingen, provided inspiration and learning opportunities that allowed my ideas to mature.

I am particularly indebted to Trevor Stack and the whole team at the Centre for Citizenship, Civil Society, and Rule of Law at the University of Aberdeen, as well as to Thomas Kirsch, Mirco Göpfert, and others at the University of Konstanz, where I presented early versions of a couple of chapters.

My gratitude and thanks go to David Lewis, Mark Schuller, and the two blind reviewers who provided valuable and thorough feedback that helped improve this book.

These years have been challenging. I would not have been able to finish this book without the support, patience, and inspiration of my wife, Monika Class. Thank you.

CIVIL BECOMINGS

Introduction

Alternative Political Engagements

Is it better to pretend that you keep the agroindustry expansion on the side, and throw rocks? Resist? Or is it better to co-opt it and to see what happens?

—Nathan, senior researcher at IMA

WHEN I ARRIVED in Belém in September 2004, Renata welcomed me and gave me an introduction to the Instituto do Meio Ambiente da Amazônia (IMA),[1] one of the leading nongovernmental organizations (NGOs) working within a socioenvironmentalism framework in Brazil. To this point, our only contact was through email while I negotiated access to the NGO for my research. As IMA's acting executive director, as well as one of its researchers, Renata gave me a tour of the NGO's headquarters, based in a French-style mansion in the city center. As she went around the house knocking on doors to introduce me to dozens of workers, technicians, and researchers in secluded air-conditioned rooms, she explained some of the main projects of the organization. When we sat down in her office, she warned me: "First of all, I must clarify: IMA is not an activist organization; we are researchers and use the knowledge we produce to find solutions to problems." She was adamant that I did not misunderstand IMA's mission, and she was also keen to find out whether I was hypercritical of NGOs. I thanked her for her clarity and explained again that I was interested in studying advocacy networks in the Brazilian Amazon through one single NGO. I did not think much of her concern then, but I came to understand it later on from comments by members of foreign aid agencies and other NGOs. They were all addressing a significant underlying tension in the region between radicals and reformers.

---

1. The names of my main interlocutors and of the key NGOs portrayed have been changed.

One arena where such tension was present was in the uses of forest land. While radical activists rejected any type of large-scale agrobusiness in the Amazon, preferring protected areas, extractive reserves, or at most small-scale agriculture, groups like IMA sought compromises that would allow for a balance between systems. Radicals were also suspicious of NGOs, while reformers used these as diplomatic tools. Nathan, an American senior scientist who was among IMA's founders and most active members, articulated his reformer credentials bluntly while I was staying in a research camp in Mato Grosso: "Is it better to pretend that you keep the agroindustry expansion on the side, and throw rocks? Resist? Or is it better to co-opt it and to see what happens? For example, small-scale family agriculture, [what] can [they] get out of this new phenomenon? There is potentially a lot of money that can be given to smallholders, ridding themselves of land that is bad for family agriculture so they can get better land for family agriculture. There are natural synergies out there with soy farmers wanting flat land with no water, and farmers wanting hilly land with streams. And such is the natural division of the landscape." Nathan's pragmatism was not shared across IMA, but it informed the group's ethos of avoiding a confrontational stance and seeking compromises. This was valid as much with agroindustry as with radical environmentalists. IMA collaborated with large agribusinesses, but it also took part in denunciations of large-scale illegal deforestation alongside radical groups.

A few months before, on the other side of the Atlantic, in Barcelona, I had joined Socis de la Terra (SdT), the local chapter of a large NGO I will call Partners of the Earth (PoE). I had chosen the Catalan capital because of its well-known status as a hub for activists and advocacy, which is strongly influenced by the region's historic progressivism combined with a nationalist agenda. SdT was my choice because PoE already worked as a network connecting its local chapters. Víctor, who led SdT, allowed me access to the group and its meetings on the condition that I work as a volunteer for the organization. For this reason, I got involved as SdT's representative in the preparatory meetings of the Mediterranean Social Forum (Fòrum Social Mediterrani, FSMed). This appointment allowed me firsthand access to the entanglements of activism and advocacy in Barcelona. From my arrival, Víctor introduced himself as a free-thinking left-winger with a hippie past. He argued that working in an NGO allowed him to be critical of the system while managing to make a living. When I arrived, SdT used an office space within a large space someone rented out to different NGOs. A terrace outside this shared office overlooked a cooperative bar where we would sometimes get a beer after a day's work, where one needed to be a member to be able to order drinks. Such arrangements are common in the city, as in the rest of Catalonia. They are the practical results of ideological convictions respecting the value of common action.

Among my responsibilities for the organization of the FSMed was to produce graphic designs for T-shirts, pins, and stickers. That is how I found out that there is a graphic arts cooperative that offered very low prices for regional movements or campaigns. The alternative milieu is therefore not limited to the sizeable number of activist groups and NGOs that are dynamically involved in public life in the region, but also encompasses other groupings—such as private enterprises—that together function as a type of infrastructure that provides services for all involved.

In both cases, Brazil and Barcelona, advocacy networks followed organizational arrangements stemming from years of activism and advocacy. Both settings also reflect the urgency of demands and difficulty of success that advocacy networks live with. As I write these lines, the Brazilian Amazon faces the threat of a regime that bluntly promotes mass deforestation to increase agroindustrial production. After decades of campaigns and negotiations, as well as scandals and diplomacy, advocacy networks had achieved governmental protection of massive swaths of land, including for indigenous peoples and for sustainable production (known as "extractive reserves"). Many of these hard-won achievements are nevertheless threatened by a new vision from president Jair Bolsonaro. The overall consensus among environmental NGOs and scientific institutions worldwide is that such destruction does not only threaten the region and its ecosystems, but also global weather patterns and biodiversity. In great part, the international reaction against Bolsonaro's plans for the Amazon has been informed by the same or similar advocacy networks that I studied. In these, IMA continues to carry out research that provides information to better understand the region's ecosystems and risks.

In the case of Barcelona, Catalan nationalism has continued to gain force in recent years, overshadowing the progressive cosmopolitan agendas with which some groups sought to justify the region's leadership in the Mediterranean. Long before the events that were named "Arab Spring," the FSMed had been thought of as a space to allow for activist and civil society groups from Northern Africa, the Middle East, and the Balkans to come together to debate and learn from their plights. Although organizers had sought to make it a periodic gathering, the first FSMed event was to be its last, in great part due to the overwhelming Catalan nationalism that eroded trust among participant organizations. The problems that nowadays characterize both sites are thus signs of our times: nationalism and a disregard for scientific knowledge regarding environmental matters. It is perhaps useful to understand some of the tensions and negotiations faced by activists and advocates who have for decades grappled with these issues.

In this book, I argue that through advocacy networks, member organizations seek to establish themselves as legitimate civil society actors who can and should deal with issues that concern social collectives. Advocacy networks are not merely the sum of their parts (NGOs, social movements, and other groups), but something new. In their collective endeavors, I contend, they go through what I term *civil becomings*. They do so through three processes: (a) a web of communication through which they can make decisions and act, (b) an interweaving collection of public performances, and (c) a constant balancing act between vernacular and cosmopolitan values. The first (a) refers to the plural character of networks, which requires a series of multilayered communication strategies. It also raises the question about the agentic potential of networks, which I call "entangled agency." The second (b) consists of a series of enactments through which network members carry out what I have termed "alternative performative politics." The third (c) refers to the manner in which advocacy networks deal with contrasting moral frameworks to pursue collectively driven agendas, and how this shapes their political agendas.

There are vast bodies of literature dedicated to activism (e.g., mobilization, social movements) and advocacy (e.g., NGOs, third sector, or civil society), but few publications seek to capture the complex interweaving of the two. Both fields explore fascinating developments of our times and provide narratives that help understand political maneuvers that may or may not influence governments and/or corporations. This book is an attempt to contribute to a third field, one focused on analyzing the frequent interactions between activists and advocates. The networked bureaucracies are formed by a wide variety of groups (NGOs, social movements, among others) that are not impervious to one another. Activists may choose to work for an NGO and in their free time take part in marches or interventions organized by social movements. Or, people may move freely between those groups and others, including government offices, in what is known as "revolving doors" between the third sector and the state (Lewis 2008). An envoy of the Environment Ministry to the Amazon region, who had worked for decades in environmental civil society groups in various capacities, explained to me that "when the Workers' Party won the federal elections, many of us in the [environmental] movement joined various government offices." This and further developments in the Amazon make for a more complex dynamism that defies static classifications. Networks between NGOs, social movements, and other groups work differently when they have allies in some areas of government. Such alliances may rely on personal contacts, similar backgrounds, or shared outlooks.

The analyses in this book are anthropological, informed by ethnographic research in both field sites. It is thus an ethnography of emerging forms of transnational political engagements. In particular, I study forms of collaboration that take place among activists and advocates—whom I also refer to as radicals and reformers—both of whom escape easy categorization. Advocacy networks are collectives where individuals, organizations, and institutions collaborate. In doing so, they not only carry out work for their stated aims but also embody their vision of "organized civil society" at both transnational and local levels. Rather than seek, myself, to define civil society as a concept, I focus on the way in which those involved carefully construct it through the collective efforts by which they address specific political goals in cooperation or in conflict with governments or international agencies. A key part of their work consists of their claiming a legitimacy that is attached to the idea of organized civil society as a collaborative form of alternative politics. In these efforts, organizational innovation goes hand in hand with a willingness to collaborate across different styles of action or ideals. The value of using an ethnographic approach to disentangle these issues comes from a deep immersion in the everyday workings of advocacy networks. In each of the two field sites, I followed a network from within a single NGO (IMA in Brazil, and SdT in Barcelona). This allowed me to get to know the everyday practices with which network members dealt to work for the joint aims and projects. It also provided valuable insights into the contexts of each NGO, in terms of place and in its sociocultural milieu.

My fieldwork consisted of sixteen months, divided between Brazil and Barcelona in 2004 and 2005, in which I carried out participant observation, interviews, documentary research (of texts produced by NGOs, governments, and about them), and photographic registry. As a volunteer in the FSMed, I participated in numerous organizational meetings, but also other events such as marches and seminars. In Brazil, I joined numerous meetings with IMA researchers and technicians and visited some of their projects. In total, I carried out fifty-two in-depth interviews with activists and advocates whose work was central to the analyzed networks. I also carried out forty-six semistructured interviews with various NGO workers, project beneficiaries, neighbors, activists, and with some people attending their events. While I did record the interviews, I only took notes in the meetings. The transcripts of the interview tapes are therefore closer to the literal utterances of interviewees. The transcripts of meetings and informal encounters are interpretations of my notes, which follow as closely as possible what took place but remain nevertheless edited. This method adheres to Green, Franquiz, and Dixon's view of transcription as a situated act (Green, Franquiz, and Dixon 1997). Among the documents I analyzed were dozens of printed materials from NGOs and activist groups, from international organizations and government offices that collaborated with them, and from print and online media about their work or ideas.

My interest in advocacy networks stemmed from the time I worked as a journalist, before my graduate studies in anthropology. I remember having to write and edit various texts regarding the 1999 protests against the World Trade Organization (WTO) in Seattle, and being amazed at the coordination so many activists achieved with such limited technological means (Eagleton-Pierce 2001). During my master's studies, I decided to analyze the World Social Forum (WSF) anthropologically, which resulted in my first monograph (Acosta 2009). For my doctoral degree, I originally sought to compare two sets of advocacy networks devoted to environmental issues: one in the Amazon and another in the Mediterranean. The contrast in regions was deliberate; I was striving to avoid a single subject (e.g., rain forest ecosystem protection) and a single politico-historical context (the Mediterranean). Additionally, I was seeking an overview of complex forms of negotiation. My choice of both areas was due to their relevance for international relations and to my aim of contrasting situations in the Global North and South. Both countries, Brazil and Spain, had also relatively recently established democratic regimes after decades under military dictatorship in the twentieth century. This provided civil society organizations with an impetus to insist on their role in upkeeping democracy. In each place, I studied networks from within one NGO to follow closely the quotidian work carried out by people dedicated professionally to their causes. In the Amazon, I was based in Belém, Brazil, and in the Mediterranean, in Barcelona, Spain. As happens often with these projects, circumstances somewhat altered the path I had imagined. The most significant change was in Barcelona, where I ended up closely following the organizational process through which a group of activists and advocates brought about a meeting called the Mediterranean Social Forum (FSMed) because the NGO I had initially selected did not take an active part in environmental networks. The result is a disparity between both field sites in that the FSMed does not comprise only environmental movements but is rather an umbrella organization including a wide array of groups. It is basically a regional edition of the World Social Forum, an alliance of activists and advocates who seek to strengthen progressive civil society projects in annual meetings. I believe the resulting contrast is actually a welcome disparity that allows for interesting analyses between networks dealing with territorial and environmental matters and those seeking an increased dialogue between independent groups. In both cases, networks are established to work toward bringing about desired futures.

Activists and advocates in both sites constantly asked me about my opinions. They were seeking clarity about my research but also about my own convictions and ideas. I gladly shared my views with them, and some of my initial reflections. I also told them I was sympathetic to their causes but nevertheless sought to make critical analyses of their endeavors. I noticed how me being a brown Mexican pursuing a doctoral degree in anthropology in the United Kingdom stimulated contrasting responses from them. Because IMA is a science-focused NGO, its research staff—both Brazilian and American—engaged with me with scholarly curiosity. Several IMA researchers—including Renata—had studied in the United Kingdom. For other Brazilians in IMA and other groups, I was an "honorary Brazilian," as I could pass as a Brazilian, and also because Mexico held a special place in their soccer hearts (it is where their team won the World Cup in 1970). In Barcelona, among labor union radicals, my Mexican nationality and my brown skin placed me in a position that deserved sympathy and solidarity (albeit in a somewhat condescending way). For many of them in both sites, it was unexpected that I was not myself an activist, as it seemed to be what they were used to. But they nevertheless respected my work and were extremely helpful along the way.

The Transnational Politic

I refuse to endorse the romanticism of scholars who, on the one hand, celebrate the formation of transnational activism while, on the other hand, criticize the world of professional nongovernmental organizations (NGOs) as merely a neoliberal stratagem of power management (Kamat 2004; Tembo 2003; Graeber 2009). Instead, I focus on the relations *between* activist movements and NGOs and especially the *networks* they form. These complex bundles of groups, which have varying degrees of formality and cohesion, are achieving numerous changes around the world—in local and international policies as well as in people's quotidian practices. Some are established institutions that dedicate considerable effort to diplomatic affairs with the aim of modifying national and international legislation and policies. Others are temporary assemblages who gather to achieve a collective goal that all participant groups and individuals agree on addressing. My argument is that the complex interweaving of so many groups into unified fronts works to catalyze wider moral debates where local principles are in flux along with so-called cosmopolitan values.

Sonia Alvarez and her coeditors of *Beyond Civil Society* (2017) argue for a distinction between what they call a hegemonic "Civil Society Agenda" and "uncivic activism." The first would be the Euro-American-funded milieu of NGOs that offers a space for "permitted forms of participation," while the second is the "actual type of mobilizations within the larger social movement field" (Alvarez et al. 2017, xi). While I acknowledge that funding and close accompaniment of NGOs, foundations, and other institutions based in Europe or the United States may constitute an "agenda" that may be foreign to local groups' interests, I argue that these efforts also coexist with what these authors call "uncivic activism" without actually affecting its processes. The distinction they propose runs the risk of reproducing dichotomies that do not help to disentangle a more complex interweaving of activism and advocacy. Alvarez and coeditors claim to avoid a simplistic dichotomy regarding threats to or advances of democracy by civic or uncivic political action through the nuanced categories of what is permitted and not permitted, or the acceptable and unacceptable, in protests and alternative political action. My analyses in this volume point to the fact that the same actors may carry out both types of political action, perhaps in different settings and contexts, or also, groups following contrasting strategies may collaborate and seek legitimacy in the eyes of government officials and the general population. The key factor is that the groups I study collaborate in networks, which are not bound by the same criteria as the groups that form them, either NGOs, social movements, or others. This means their work cannot easily fit within one of the two categories proposed (permitted or not permitted forms of protest).

The plural social assemblages I refer to, furthermore, differ in a number of ways from movements with a single structure and identity. By joining forces and ideas, activists and advocates seek to establish a common moral ground. They strive to do this by claiming to embody an organized civil society that, through its activism and participation in public affairs, earns recognition from other power actors (mostly governments and international organizations, but also industry and business associations). To a large degree, activists who participate in networks seek to legitimize their work for their constant collective negotiations. Their work is, crucially, performed in various settings in order to display the challenges they address as well as the potential solutions they may propose. Such procedural moral work must address multiple cultural, material, and socioeconomic contexts. The distinct achievements of many such networks is not simply due to the influence of large and affluent NGOs imposing their agenda on the rest of the network's members. Instead, these networks are often conflict-laden, which may explore contrasting—and sometimes contradictory—methods to try to reach the results they look for. It may well be the case that within the network some groups act in an "uncivic" manner to get their point across. The resulting negotiations therefore entail an enacted conciliation between cosmopolitan and vernacular values.

In recent years, the concept of "network" has become a key looking glass through which scholars from social sciences and humanities seek to view social interactions. Its descriptive quality offers a portrayal of forms of association and relation that is less binding than preceding grand concepts of "structure" or "function." To some extent, there is now a "network approach" that seeks to contribute more thorough understandings of sociality in its material context, and to do so in more depth than is found in previous theorizations on "structural" or "functional" sociocultural arrangements. Often referred to within a "poststructuralist" turn, such an approach can be said to be closely related to *agencement* (in French), or "assemblage" (in English), which Deleuze referred to as a "multiplicity" that exists in a "co-functioning" through the relations between its elements (Deleuze and Parnet 1987, 69). Deleuze's reflections emphasize the heterogeneity of the assemblage, as DeLanda explains: "unlike organic totalities, the parts of an assemblage do not form a seamless whole" (DeLanda 2006, 4). These theoretical reflections draw much from recent thinking in the natural sciences, especially from physics and ecology. Two major lines of thought have been particularly influential in anthropology. On the one hand, Latour's actor-network-theory (ANT) provides a practical "tracing of associations" (Latour 2005, 5). On the other, Ingold's work on correspondences between life forms and objects argues that the world is actually "open," meaning that no boundaries exist between things: "no insides or outsides, only comings and goings" (Ingold 2008, 1801). These conceptual approaches to relations in materiality and existence have marked a turning point in addressing changes, phenomena, and sociality. Attention to affect, for example, is not limited to emotions, but rather explores the different influences and effects that things and life forms have on each other (Navaro-Yashin 2009). It has also provided fertile ground for analyses of political performances (Juris 2008b). More broadly, its influence is felt as a disruption of understandings of the social as well as of humanity's pedestal in scholarship so far.

Over the last two centuries, anthropologists have developed theories to convey the idea of a social system by considering arrangements as following structural and/or functional characteristics (Layton 2012). These developments have served as master narratives for a variety of analyses by which societies could be dissected. Careful attention to kinship relations and other social interactions nevertheless led to the development of wider network analyses. Political anthropologists from the Manchester school, for example, focused on ceremonial and ritual power—like through courts—and questions about state-based and stateless societies. Studies of colonialism considering the view of a capitalist world system and its effects led anthropologists like Mintz and Wolf to widen the net of meaningful interactions and constellations across vast distances (Wolf 2010; Mintz 1986). All of these developments helped shape anthropologists' contributions to the current "poststructural" age in which the concept of the "network" has become a useful analytic tool to emphasize social connections without a predetermined arrangement. Human collaboration, trade, contact, and cross-fertilization (of ideas, imaginings, language) have thus been approached with a particular interest in mapping the relations between them in order to grasp the flows of objects and the ideas they enable. Thus, the notion of networks as fundamental to mapping human endeavors is not really new in anthropology, particularly as regards conceptualizations of exchange. Almost a century ago, Malinowski identified how certain objects laden with cultural value traveled great distances among Pacific islanders in ceremonial exchanges (Malinowski 1972). Nevertheless, the current attention across disciplines provides fertile ground for the dense ethnographic analyses that anthropology can provide in theorizations of networks. Latour's actor-network-theory, for example, is a very anthropological take on representations of links and flows as well as what these mean for transformations and perpetuations of social arrangements (Latour 2005).

These reflections serve as a basis for scrutinizing the organizational arrangements that are analyzed in this volume and that comprise individuals and groups of self-proclaimed activists, advocates, development workers, and others. Advocacy networks are therefore social assemblages of people striving to have an influence on the way states are governed and international organizations are run. This book presents a study of what I call "alternative politics" as it is shaped by collectives that work predominantly outside of state government bureaucracies but that seek to influence their policies and projects. The broad networks established by a wide array of informal movements and institutionalized NGOs present an example of political negotiations through their constant dealings with difference. Their efforts to influence state bureaucracies in order to modify laws and projects testify to a type of entanglement that defines them: being nongovernmental but aiming to influence governments. The network arrangement they follow allows for different scales of action and involvement. Small groups concerned with local issues usually join efforts with other collectives working within the same locality. They may then team up with those of other localities to push for regional changes, then scale up to national level, and sometimes even into the international arena. The result is a mesh of groupings that provide renewed opportunities to imagine communities through collectively debating common interests and aims. It is not a process that draws on a sense of territoriality or identity, but rather one that reproduces sociability. These networks thus configure unprecedented forms of political consciousness. They become not so much a "moral compass" as a "political compass" seeking to shape agendas and debates in communities that may be defined by territory, ethnicity, or established boundaries.

Scholars of various disciplines have explored such organizational arrangements and their political roles. Approaches using categories and theorizations developed for social movements, civil society, or the third sector have informed a range of studies that consider various constellations of groups and individuals. Neoinstitutionalism and policy networks are other approaches that acknowledge and theorize the increasing plurality of participating collectives in public affairs. The analysis I present in this volume is an ethnographic examination of "transnational advocacy networks" as defined by international relations scholars Keck and Sikkink (1998a). These authors clarified that they preferred to call them networks, rather than coalitions, movements, or representatives of civil society, "to evoke the structured and structuring dimension in the actions of these complex agents, who not only participate in new areas of politics but also shape them" (Keck and Sikkink 1998b, 4). They thus alluded to sociocultural processes through which agreed-upon behaviors and norms shape social arrangements. At the turn of the millennium, Keck and Sikkink argued that scholars had been slow to recognize a growing phenomenon, in which "individuals and organizations have consciously formed and named transnational networks, developed and shared networking strategies and techniques, and assessed the advantages and limits of this kind of activity" (Keck and Sikkink 1998b, 4). Since then, scholars and policy makers in international relations have increasingly considered such networks to be relevant stakeholders alongside state governments (Macdonald 2008; Josselin and Wallace 2001). Historical analyses have led some authors to suggest that such networks are actually constructing a "world culture" of shared principles (Boli and Thomas 1999), especially regarding human rights (Khagram, Riker, and Sikkink 2002). Political analyses of the wide range of possibilities and implications that are opened up by these networks have expressed caution about the potential misuses of their power (DeMars 2005).

Civil Becomings

The concept of civil becomings is used in this volume to represent the process through which advocacy networks construct the legitimacy they seek as organized civil society. I use the concept of civil society not as a given but as the aspiration of various groups who use intellectual debates as an inspiration for seeking a role in how the state of which they form part is run. In Jeffrey Alexander's words, "Civil society is a project" (2006, 551). Contrary to political scientists and philosophers (Ginsborg 2008, 50), I seek not a normative stand on civil society but an analytic consideration of claims and practices. In seeking legitimacy as civil society actors, activists and advocates try to use the legitimacy that the concept of civil society entails without needing to make references to any theoretical underpinnings. Klaus Eder suggests that groups "imagine," "practice," and "stage" their conception of what it means to be civil society organizations, usually understood as "a site where people construct a social bond which demands solidarity from all to realize the common good" (Eder 2009, 24).

For Hann, "political anthropology could adapt the term civil society to open up a kind of comparative political philosophy, concerned with all the diverse ideas and moralities that inspire cohesion and trust in human communities" (Hann 1996, 22). If this is possible at all, however, there needs first to be a critical assessment of the implications of the term. Comaroff and Comaroff have considered the implications of the term for Africa. They argue that it has become a myth of diffusion from Enlightenment origins, through theoretical considerations, to struggles against totalitarianism, and to a reification through a remembering with which Euro-American intellectuals have sought to establish a genealogy of civicness (Comaroff and Comaroff 1999, 4–5). Comaroff and Comaroff nevertheless insist that it is among those who write about "civil society" or claim to put it in practice that the concept—"the irony of rendering singular an obdurate plural" (Comaroff and Comaroff 1999, 6)—is brought to life. "Civil society" therefore means different things to different people. To neoconservatives, it describes the libertarian ideal of a depoliticized society escaping the control of the state; to post-Marxists, it is a way in which people achieve leading a life free from the influence of market or state; and to those claiming a postideological stance, "it speaks of moral community and unfettered, self-sustaining public sphere" (Comaroff and Comaroff 1999, 6). Outside of Euro-America, its history and enthusiastic backing by international institutions and numerous affluent states make it a sphere of contention in itself because of the distrust of developmental agendas that may erode local agencies.

In some cases, the idea that organized civil society equates with democracy has been taken to extremes, as occurred with some academic analyses using the number of NGOs in a given country as a criterion for evaluating the quality of democracy (Sampson 1996). Some assessments of the formation of civil society urge caution, contending that it may not be the most effective way to consolidate democracy. For example, Encarnación argues that "the prevailing view of civil society as an infallible democratic miracle worker is arguably the most problematic conventional wisdom to be attached to civil society in the last few years. It amounts to a myth" (Encarnación 2003, 4). He goes on to quote a study of nineteenth-century Europe that argues that the countries with the densest civil societies, such as Germany and Italy, struggled to develop enduring institutions in the twentieth century (Encarnación 2003, 5; Bermeo 2000). Encarnación also points out that there is a paradox in newly democratic societies, where the quality of governance "has deteriorated with the progression of the institutionalization of democratic practices such as free elections, leading to widespread repudiation of political institutions" (Encarnación 2003, 175). This points to nonlinear developments, which are more relevant than ever to understanding the complex processes of political negotiation we are witnessing worldwide.

I use *civil becomings* to highlight the open-ended processual character of advocacy networks' endeavor of building their own legitimacy through their practices. It draws from Ingold's opinion that we should no longer refer to ourselves as "human beings" but rather as "biosocial becomings." In emphasizing "beings," he argues, the idea is that it is a static form of life. But because life itself is dynamic (Zubiri 2003), we should instead seek to capture the flow of biological and social life that we are made of, "that is, not as discrete and preformed entities but as trajectories of movement and growth" (Ingold 2013, 8). In their ongoing endeavors, advocacy networks thus bring together a series of practices, ideas, and aspirations through which their members pursue an influence they could not attain by working in separate groups.

Thinking of *civil becomings* as a process brings temporality to mind. Analyses of some collective efforts have recently used the label of *prefigurative politics* to describe the manner in which activists seek to establish the future they envision in their own practices (Razsa and Kurnik 2012). Prefigurative politics, however, are commonly claimed by radical left movements who prefer to set up practices that are more in line with how they think the world should function (Thorkelson 2016). This is not the case with advocacy networks. Both sets of advocacy networks analyzed in this volume are not radical; that is, they do not follow an agenda that seeks a deep change. I believe this is the case in most advocacy networks, as they bring together groups from various segments of the progressive left spectrum. As spaces, advocacy networks are closer to what radicals often call *reformist*, as their aim is to foster dialogues between actors with contrasting viewpoints to reach legitimate agreements. In the Brazilian Amazon, this challenge regards fundamental differences in positions on uses of territory and its biodiversity. In the Mediterranean, the challenge at the time of fieldwork was to foster encounters and dialogues between Europe, the north of Africa, and the Middle East. In both places, the networks I followed eventually fizzled out; they were, however, replaced by new networks, in which those involved applied lessons they learned in their previous networked efforts.

Instead of being a case of prefigurative politics, advocacy networks appear more as the form and content of an intensely relational character. That is, they become a *dispositif* through which subjectivities are formed. This is different from prefigurative politics, as activists and advocates do not consider their collaboration to be an example of how the future should be. They rather use the formula for its potential legitimacy in making demands or proposals. They appear to not notice that in constructing such legitimacy, they are shaping their own practices of listening to others and negotiating. This should not be understood as a purely positive process. The rift between radicals and reformists is an existential one for many of those involved, with each side claiming victories it affirmed were a direct result of its efforts. In

Brazil, environmental agendas have developed over the last decades as a result of the interplay between both radicals and reformers and government institutions (Hochstetler and Keck 2007). From what I witnessed, however, it seemed to me that all sides needed each other and collaborated in ways that they nevertheless kept somewhat hidden. As happens with other ideologies—political, religious, or identity based—the appearance of purity, or uniformity, is useful for maintaining both allegiances and a united narrative of action and belonging. The collaboration I notice, however, highlights an emerging correspondence where contrasting actors increasingly understand the interrelated character of a problem and decide to address it through a cooperative approach.

In my view, the few anthropological studies of activist or advocacy networks remain either one-sided, choosing to identify with the network causes, or too detached, paying attention to aesthetic qualities of collaborative work. In the first camp would be the work of those who study radical groups, such as the "alterglobalization" or "anti-corporate globalization" movements, from a sympathetic angle, such as Juris (2008a), Graeber (2009), and Maeckelbergh (2009). In a contrasting vein, Riles focused on networks of NGOs as they prepared to attend a conference organized by the United Nations (2000). These books seem to polarize a field that, from what I witnessed, is not so polarized.

Advocacy networks are considered here as political experiments in pluralism. The most effective networks I witnessed were those that existed in a state of anxiety—of impermanence because of the looming risk of fragmentation. This kind of tension is actually very productive. It triggers a need for ongoing negotiation that would otherwise be offset if there were an overarching agreement among all participants. From what I witnessed in the Brazilian Amazon and in the Mediterranean, the true alternative character of advocacy networks lies in their openness to a wide diversity of viewpoints and agendas that require impromptu negotiations and arrangements. It may be that the difficulty in understanding the way that networks operate lies in the language we use to describe them. It is common for analyses to stress their actor-like character (i.e., as "nonstate actors" in international relations) when in reality they are bundles of contrasting positions that collectively react. It is a case of organizational "swarm intelligence" (Rheingold 2002). While I maintain the category of advocacy network, I believe it is useful to consider advocacy networks through the lens of assemblages, as they are constituted by a multiplicity of organized collectives (DeLanda 2010a, 10).

Entangled Forms of Political Sense-Making

My focus on the agentic, performative, and moral characters of advocacy networks situates them in increasingly complex political scenarios. While environmental concerns are not limited within political borders, states' policies have disparate effects in specific situations. It is perhaps a case of organizational pluralism that requires a different approach. Seligman and Weller have argued that the more common symbolic mechanisms through which collectives used to deal with ambiguity—such as ritual or notation (the formation of categories)—have been eroded in recent times due to the increasingly mixed character of communities (Seligman and Weller 2012). They argue that there is currently a reduced presence of uniform rituals whose performativity has historically helped people enact a reifying as well as a crossing of boundaries that is essential to constructing agreed understandings of social life. These processes are straightforward when religious ceremonies are performed by collectives who can understand the various shared symbolic meanings. In increasingly multicultural settings, where cultural references are no longer shared among all, Seligman and Weller claim the need for new mechanisms that include a greater diversity of backgrounds and expectations. Advocacy networks, as spheres where activism, advocacy, or an active search for alternatives to states of affairs converge, may well provide such opportunities to bridge experiences between people with contrasting visions. That is, at least, what I noticed in the Brazilian Amazon and Barcelona.

Latour has argued that networks pose a challenge that requires understanding the flows within them as much as the organizational structure they have. A gas network, he insists, is not made up of gas, nor is a water network made up of water (Latour 2013). The pipes, stations, outlets, and other materials involved in the actual physical layout that makes up the network allow for its content and purpose to flow within it. His actor-network-theory seeks to trace the associations that link actors. He insists that "there is no society, no social realm, and no social ties, *but there exist translations between mediators that may generate traceable associations*" (Latour 2005, 108–9, emphasis in original). Marilyn Strathern complains that Latour's use of the concept of the network portrays an "endless extension and intermeshing of phenomena" (Strathern 1996, 522). She argues for the need to "cut" such networks so that there may be a possibility of portraying and analyzing them. She uses the example of the way in which anthropologists in the 1950s and 1960s studied kinship relations. If one were to register the whole network of relations of any given individual and group, these would extend endlessly. So, she adds, "It was argued that in order to create groups, for example, ramifying kin ties had to be cut through other principles of social organization" (Strathern 1996, 529). This means that a particular focus needs to exist for a researcher to select which relations will be analyzed. It is not a matter of following all traces, but of identifying those that are relevant for a particular study and examining them.

For the purpose of this volume, therefore, the networks that are traced are those that aspire to reshape political landscapes by reconfiguring decision-making processes. This takes place in the political arena, where common issues are debated, negotiated, and decided on. The groups and individuals I met in Belém and Barcelona were keenly motivated in their particular projects and in the networks they formed. Some saw their individual efforts as part of a larger movement that was directly benefiting the bigger picture. For example, in Brazil, some scientist-advocates would think of their research projects as platforms for exploring challenging questions so that they could better understand the forest at a crucial moment of its history—at a time when it is threatened by a large-scale push from industrial agribusinesses and farming. In Barcelona, labor union activists would openly refer to the World Social Forum as a process that had inspired them because of its collaborative character of grassroots decisions. Many other people I met in both sites, with very different styles of activism and advocacy, told me it was better to work together than apart.

My interest is therefore to find out how such diverse and fragile coalitions can work together. Their flimsy collaborative status appears to make them agile and swift insofar as they can improvise or react more creatively to unexpected developments. Bureaucratic institutions or organizations, which are bound by numerous rules and contracts for their internal procedures and their external links, are effective but slower to react. When they link up with other groups, the combination of skills and capacities, but crucially also of ideas and values, provides the collaborative work with an energy and dynamism that would not be possible if each group were working alone. In forming networks, therefore, they are not merely aggregating labor by adding hands to a campaign; they are reconfiguring the way in which certain issues are perceived and thought about. Thus, they are swaying subjectivity. This is a process that Hardt and Negri call "biopolitical production" (Hardt and Negri 2009, x): "Politics has probably never really been separable from the realm of needs and life, but increasingly today biopolitical production is aimed constantly at producing forms of life. Hence the utility of the term '*bio*political'" (Hardt and Negri 2009, 175). Henrietta Moore argues that the political is nowadays producing new social imaginaries that are "not so much about ideational forms or ideologies as about emergent socialities" (Moore 2011, 163). In this sense, the process of identification with an ideal is not as strong as the practices through which actors work toward specific goals. Activism and advocacy follow this way of shaping social life, through ideas, decisions, and actions.

The arena where these networks are active is local and yet also transnational, with parallel actions occurring in specific locales and in national and international spheres. Activists and advocates usually deal with state government officials in some capacity in order to achieve changes in the running of collective life, either in the form of public policies or projects. Because of such involvement of the way in which decisions are made for collectives, political anthropology frames this book. It seeks a place in what Spencer recently called the "newly emerging anthropology of the political" (Spencer 2007, 5), in which the political is understood as "a complex field of social practices, moral judgements, and imaginative possibilities" (Spencer 2007, 22). Rather than focusing on direct forms of engagement between individuals and government authorities, as the participation in political elections may be or in local forms of government, what advocacy networks do is redefine the terms of defining priorities for social life. In part, this is related to Holston's concept of "insurgent citizenship" (Holston 2008), which he used to describe the hands-on contestation of institutional inequality. His rich ethnography describes how people in São Paulo's urban periphery built their own houses and neighborhoods as a way of shaping their own belonging. The relation is limited to the performativity of joint action to give shape to desired outcomes for collective life. Holston's case is more directly linked to the sense of national belonging (citizenship) than advocacy networks usually claim to have, especially when they work in a transnational arena.

Social movement scholars have for a long time acknowledged the complex associations that occur in collective mobilizations, especially in our transnational and digital age (Edelman 2001). They usually consider social movements as constellations of individuals and groups who opt to come together for joint action. But this is where it gets fuzzy. While some scholars choose to remain with the label "social movement" when the range of organizations involved includes large NGOs, labor unions, and others, I choose to name these umbrella organizations "advocacy networks." It is not a label that its members would use, but is rather an analytic category. Advocacy is for my purposes diplomatic activism, or the search to influence collective processes that are often in the hands of governments. More recent studies of advocacy efforts have sought to define their scope, as Reid does: "Advocacy activities can include public education and influencing public opinion; research for interpreting problems and suggesting preferred solutions; constituent action and public mobilizations; agenda setting and policy design; lobbying; policy implementation, monitoring, and feedback; and election-related activity" (Reid 2000, 1). Recent attention on advocacy has come about because of the rise of such networks.

The burgeoning field of anthropological studies of NGOs is also relevant. Emerging from critical development studies, scholarly interest in NGOs has sprung up in an attempt to capture the highly diverse settings and projects where such groupings work (Lewis and Opoku-Mensah 2006; Mertz and Timmer 2010). NGOs are relatively flexible institutions that are able to adapt quickly to changing situations—as, for example, in reactions to natural disasters, health hazards, or armed conflicts. But they have also been identified as problematic groups who perpetuate privileges for a small elite while hiding behind a rhetoric of democracy and "good governance" (Abramson 1999). For this reason, DeMars calls them "wild cards in world politics" (DeMars 2005). Despite the extensive analyses showing how NGOs can easily fall prey to a neoliberal system of privatizing public services and perpetuating audit cultures, there are also numerous examples of "dynamics of hope" (Mertz and Timmer 2010, 175).

This book is not a starry-eyed championing of advocacy networks, but neither is it an attempt to simply cast advocacy networks in a negative light because of their unfavorable aspects. I seek to offer an analysis of what it means for people to collaborate, either with an aim to change things, to state their opposition or critiques of projects or policies, or for a wide variety of reasons. Most activists and advocates I met were passionate about their work. Some of them did it for a living, with wages from an NGO or a labor union. Others did it in their free time, seeking to make contributions to what they saw as a wider movement. The advocacy networks I refer to are thus not simply aggregations of efforts. They become something other than a sum of individual practices. The Brazilian case represents a clear manifestation of this characteristic. Donato, whom I met in Manaus in a conference of environmental NGOs, invited me to visit Chico Mendes's house in the state of Acre. After Mendes was assassinated, he became a martyr of Brazil's environmental movement. As I walked through his house and around the town of Xapuri, Donato regaled me with stories of what Chico had done and what he meant. I talked to several people who were with him in his campaigns, who were accustomed to meeting researchers and journalists asking about Mendes. He had obviously been a charismatic leader who lived by his own creed. Although rooted in the Brazilian Amazon, Mendes acknowledged that the attention and support he received from foreign organizations had been crucial for the movement he led. Some international NGOs provided grants before other groups in Brazil started helping him.

In Barcelona, the density of activism and advocacy meant that there were also plenty of antipathies across the board. These were even more extended at the Mediterranean level more broadly, which became evident as the preparations for the FSMed event progressed. Even though each cluster of groups would be extremely diverse, there was often talk of "the Greeks" or "the French" as if they were a unified front. Among the foreign factions, I heard that there was distrust against "the Catalan" groups. The leadership by

Catalan groups in the FSMed was looked on by many foreign activists with suspicion, as it was considered to be aligned with an official policy of the regional government to strengthen its influence across the Mediterranean. This view was further reinforced when the FSMed's official inauguration took place at the Palau de la Generalitat de Catalunya, which houses the Catalan government. Although the World Social Forum process requires support from local governments for the running of the event, it is not common for governments to hold an official opening ceremony at such events. In this case, the Catalan government clearly used the opportunity to position its role as an engaged mediator between European and non-European regions around the Mediterranean. The speaker, an official of the Catalan administration, openly talked about the government's support of civil society organizations.

Organization of the Book

This book is divided into four parts, each of which is guided by a concept that helps frame the analyses of advocacy networks and builds an argument that will culminate in the conclusion. Each of the parts comprise two chapters, which are themselves focused on a key concept through an ethnographic analysis of one of the two cases. The same concepts would have been relevant for both, but are explored with a single example in order to provide more depth to the analysis.

The first part, "Settings," situates each case in its own context while also laying out the crucial historical conjuncture through which the conceptual and methodological framework of activists and advocates is produced. Many advocacy networks operate in a transnational sphere, but the transnational is made up of a multiplicity of localities. In working with patterns of similarities and differences, advocacy networks thus function as learning structures that are able to transfer knowledge, people, and resources between areas to address similar problems or issues. Each of the two chapters included in this part addresses the recent history of each of the two areas where I carried out my fieldwork, Belém and Barcelona. The first tells the story of socioenvironmentalism, the Brazilian environmental movement that developed during the second half of the twentieth century, which has its roots in the region's colonial history. The second analyzes the story of Barcelona as a symbol of prosperity and contestation in the Mediterranean, especially in light of its defiant milieu, its urban model, and the abuse of both that took place at the "Forum Barcelona" during my fieldwork. In both cases, even though the situations addressed by the advocacy networks I observed were truly transnational, each network was molded by the particular contexts in which I observed it.

The second part of this book is titled "Plural Networks" to emphasize the implications of the diversity that partially defines activist and advocacy networks. Such composition is a fundamental characteristic that requires a series of negotiations between participants to decide common standpoints. This process, teeming with friction, provides these coalitions' high levels of legitimacy and expertise that are useful for negotiations with state and non-state actors alike. It is therefore an arena where conflict and diplomacy are constant practices, especially when all those involved have particular expectations for changes that may be brought about by the network's exertion. Drawing on Mouffe (2005), the sociality *within* and produced *by* the networks is here defined as agonistic, where conflicts and disagreements prevail and yet common positions are reached (albeit often reluctantly). The first chapter of this part focuses on the communicative characters of advocacy networks that forge dialogues out of dissenting views in a more haphazard way than religious or government institutions do. These apparently chaotic processes are perhaps exactly what is required for negotiations to be creative. The second chapter analyzes the agency of an advocacy network as distinct from that of its constituent parts. Agency is here considered as the capacity to make a decision and take action on it. Because of their plural character, advocacy networks are not homogeneous groups that can clearly stipulate how decisions and subsequent related action are made and taken, respectively. Thinking of a network agency, therefore, entails exploring the implications of a purposeful social assemblage, in Deleuze's terms (DeLanda 2010b).

The third part of this book is called "Alternative Performative Politics," as it focuses on the performative character of activists' and advocates' alternative politics. It is crucial to point out here that the networks studied in this volume are considered not only as sites of protest or public objection but also crucially of diplomacy and production of knowledge. This in itself widens the focus to include not only marches or open defiance through what social movement scholars refer to as repertoires of contention (Tarrow 1998; Traugot 1994) but also practices such as meetings and scientific research. The first chapter of this part analyzes the counterpoint between marches and meetings in the FSMed, which mirrors the web's collective public and private forms of contestation. The second chapter focuses on the production of scientific knowledge in Brazil as an example of a collective endeavor of advocacy networks that is deemed necessary to effect an influence in policies regarding the Amazon.

The fourth and final part is called "Informed Aspirations" because it explores the implications of the work of advocacy networks in terms of morality and democracy. Mobilization requires individual commitments within collective assemblages, which is at the base of negotiations over morality and democracy. The first chapter is dedicated to exploring the moral entanglements that are produced by networked advocacy. Anthropology approaches the study of morality as a collective construction of what is right and wrong, where the legitimacy it requires to become a reference for behavior and thought is based on a sense of community. Although activists and advocates do generate a type of enclosed sociality among them, their aim is to bring their agendas of change to wider social assemblages. These in turn often do not share that sense of belonging, which is often linked to moral communities. So what is being constructed with advocacy networks is a more fluid sense of relational morality that can be adapted to distinct contexts and histories without a preexisting community. Activists confront the challenges such construction entails every day by making decisions that seem best, at least in the moment in which they are made. These often occur among those involved in cultural translations, and often amid fractious misunderstandings or distrust. And yet, such moral entanglements grow in complexity and reach. How they can change local configurations remains to be seen. The second chapter analyzes the way in which advocacy networks function within a framework of democracy as well as how they work to further its cause. This is part of the purpose of the idea of civil becomings: to reimagine an improved capacity scaffolding for good governance in order to allow for collective decision-making mechanisms—that is, to help sustain and improve the democratic process that allows for activism and advocacy in the first place.

This book investigates the nascent ceremonies of legitimation shaped by the actors involved in advocacy networks. Just as Latour speaks of scientific practice as a multilayered process through which those who take part legitimize each other as scientists, so do activists constantly recognize and legitimize each other's role not in specific campaigns but in the collective endeavor of a political community guided by democratic ideals. I hope to provide insight into the "ceremonial grammar"—the "forms of agreement in the way we 'dance' together, forms of agreement in ceremonial participation" (James 2003, 92)—that these webs of purposeful action are seeking. The basic premise of this volume is that by engaging in political activities across state and cultural borders, these groups are reconfiguring the forms through which individuals engage with polities or political communities. In their engagement with political issues and management of expectations, they appear to follow a similar path to that of political parties when they were first formed. Political parties arose as networks of individuals with similar interests seeking to enter government institutions (Ostrogorski 1974). Like political parties and government officials, NGOs and social movements claim to work for the common good. Advocacy networks around the world are helping local groups shape the political agendas of their state governments. Many people partake in issues that are far from their own localities either by donating money or investing time in seeking a potential solution to a specific problem. They thus enter into a reciprocal relation that establishes a basic social bond or sense of solidarity.

With hundreds of thousands of NGOs and social movements alive and active around the world, a better understanding of their intertwined significance is crucial to an enhanced comprehension of their political significance. This anthropological study contributes to such a scholarly aim that encompasses cultural configurations and collective behavior. The value of ethnographic research lies in its capacity to tell a story of how people live and make sense of their actions. The purpose is not to offer a typology of movements or organizations and their implications, as this would imply established categories through which such actions are carried out. I take to heart Bourdieu's critique of a superficial understanding of kinship: "When the anthropologist treats native kinship terminology as a closed, coherent system of purely logical relationships, defined once and for all by the implicit axiomatics of a cultural tradition, he prohibits himself from apprehending the different practical functions of the kinship terms and relations which he unwittingly brackets; and by the same token he prohibits himself from grasping the epistemological status of a practice which, like his own, presupposes and consecrates neutralization of the practical functions of those terms and relationships" (Bourdieu 1995, 37). The key to understanding kinship, he argues, is in grasping them as "practical" relationships: "practical because continuously practised, kept up, and cultivated—in the same way as the geometrical space of a map, an imaginary representation of all theoretically possible roads and routes, is opposed to the network of beaten tracks, of paths made ever more practicable by constant use" (Bourdieu 1995, 37–38). Thus, what is sought is to follow the situated practices and their contexts in order to better understand networked activism and advocacy. I choose to view the process in which all these practices are articulated as continuously seeking to define priorities for local, regional, or larger populations, thus legitimizing the participating groups and individuals as key social actors in an ongoing process of civil becomings.

PART ONE

Settings

1

The Brazilian Amazon and Its Socioenvironmentalist Movement

We want to build a network that becomes a reference in finding a balance between production and conservation. For Brazil itself, because the Xingu River is not the only one in the country with problems. It is in the interest of all our communities for us to work together.

—Marcelo, from the Instituto Socioambiental, at the launch of the Y'Ikatu Xingu campaign to protect the springs that feed the Xingu River

IT IS A very hot day in Canarana, a small city in the state of Mato Grosso, in Brazil. Dozens of people are gathered at an auditorium for the initial talk of a conference that will run for three days. It is October 2004, and some of those present have never set foot in this type of event. There are federal, state, and local government officials, representatives of indigenous communities, small farmers, schoolteachers, academics, agribusiness employees, and many NGO personnel, some of whom convened this meeting. The purpose of the gathering is to launch a campaign to protect the springs from which the Xingu River (a tributary of the Amazon River) rises. All those attending have an interest in the issue and listen attentively to Marcelo, the first speaker and a member of one of the convening NGOs, the Socioenvironmental Institute (Instituto Socioambiental, ISA). He insists on the need for dialogue among all those present because "it is in the interest of all our communities for us to work together." On a giant screen behind him, attendees view a rotating image of Earth as seen from outer space. The view suddenly zooms in on Brazil, and then onto the area around the Xingu basin, where the image freezes and the title of the meeting appears (figure 1.1). Marcelo's talk, like those of many other speakers over the next couple of days, is full of references to Brazil. The

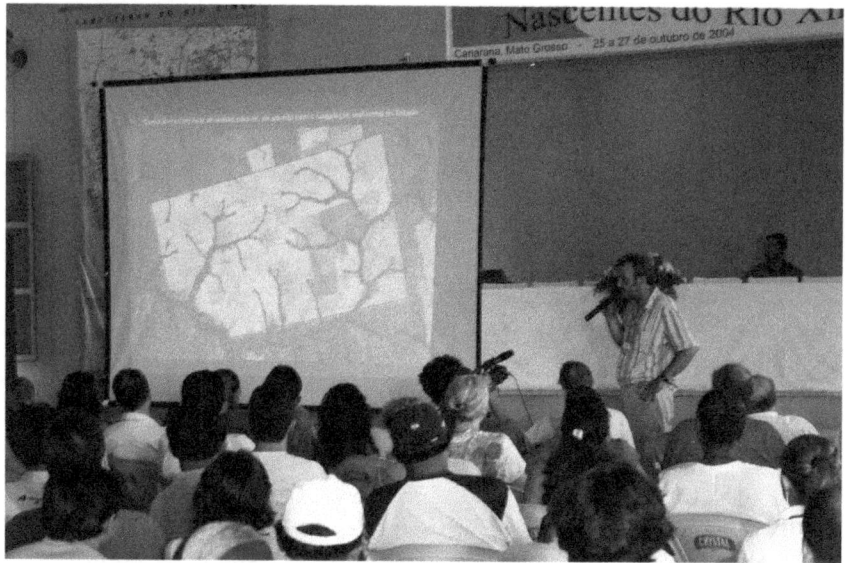

Figure 1.1. Marcelo, a senior advocate at the Instituto Socioambiental, welcomes participants to the three-day meeting to launch the campaign Y'Ikatu Xingu to protect the springs of the Xingu River in Canarana, Mato Grosso. (Raúl Acosta)

central theme of the gathering is clearly environmental and, as he says repeatedly, "for the good of the country." The benefits of the project about which he is speaking will be, he explains, social, economic, and political, and they will be for all Brazilians. His nationalistic rhetoric is particularly well received because both of the convening NGOs—Instituto Socioambiental and Instituto do Meio Ambiente da Amazônia—are Brazilian organizations. What he fails to mention, however, is that both NGOs belong to several transnational networks and are funded by and work closely with international foundations and agencies.

In this chapter, I describe the setting of the advocacy networks I studied in my fieldwork in Brazil. I do so by telling the story of socioenvironmentalism, a vernacular approach that combines protecting biodiversity while also addressing social justice matters, through the eyes and experiences of activists and advocates I met. During my time in Brazil, I visited several places in the Brazilian Amazon where Instituto do Meio Ambiente da Amazônia (IMA) led or participated in projects with other organizations. This allowed me to better understand the mosaic of interests and concerns that lie at the base of socioenvironmentalism. For many of those I spent my time with, the ecosystems they inhabit are part and parcel of who they are. Many of them thus incorporate socioenvironmentalism as a way of life, which is constantly under threat by concerted extractivist efforts of miners, or large agribusiness projects of soy producers or large-scale farmers. Among the places I visited was Xapuri, in the state of Acre, where Chico Mendes lived and led his efforts not only against deforestation but also for the rights of forest dwellers. As I will explain, Mendes formed a crucial alliance between rubber tappers, indigenous peoples, and small-scale farmers that was called Peoples of the Forest. In doing so, he and his collaborators experimented with what has become a burgeoning site of intercultural dialogues and alternative politics. Their influence has been considerable, as has been also the reaction against it.

A Forest of Alternative Politics

Advocacy and activism are common alternative political activities in the Brazilian Amazon. To achieve these activities, numerous individuals form a dense web of alternative political practice. As is also the case around the world, a recent trend among these groups is to collaborate on joint projects, that is, to form active networks. What this entails is a need for intercultural dialogues, negotiations, and actions, with at least four types of actors: first, with local communities where each organization works, identifying problems, concerns, or aspirations (e.g., indigenous groups, small farmers' unions); second, with partner organizations and local government officials, deciding and coordinating actions and campaigns; third, with foundations, international organizations, or aid agencies, designing projects, attending training sessions, and liaising with officials; and fourth, with federal government officials. In NGO-talk, all of these actors are "stakeholders" (Sachs and Rühli 2011; Friedman and Miles 2002). As was evident in the Canarana conference, to convince individuals and organizations to attend events convened by NGOs, various issues are crucial, like the legitimacy of the conveners and the aims of the gathering as well as the event's overarching framework. The framework I encountered in most events I attended in the Amazon was that of socioenvironmentalism, a particular approach to environmentalism and development that is rooted in Brazil. This chapter is focused on telling the story of this approach as a necessary and distinctive context of IMA's activities.

In all meetings I attended in Brazil where NGOs and social movements played key roles, such as the one in Canarana, the word "socioenvironmental" was commonly used by activists and advocates. It stands for the vernacular approach to environmentalism and development, "which link[s] environmental degradation and social justice" (Hochstetler and Keck 2007, 98). It sprung from a dense milieu of "domestic Brazilian thinking on environmental issues and national economic priorities and the pioneering efforts on the subnational level to begin to reconcile economic development and sustainability" (Alonso and Clémençon 2010, 250). Its roots, however, lie in Brazil's historical sense-making of its vast territory (Pádua 2012, 2013). It represents a particular challenge or alternative to the two perspectives that dominate economic and social policies in the region: developmentalism and conservation. Many environmentalists and human rights advocates have considered these the main threats to the region since the 1980s.

Developmentalism represents the effort by policy makers to prioritize economic development in the use and organization of territories and populations (Watts 1993). In Brazil, it was applied by hurriedly clearing forested areas for soy production, farming, timber extraction, and mining as well as to build roads and infrastructure in order to bring larger populations to the Amazon that would theoretically allow more economic activity (Grosfoguel 2000; Sikkink 1991). In this view, the need to protect the forest was not only irrelevant but indeed was a threat to the right of Brazil to improve its economy. "Europe and the United States achieved their economic maturity by cutting down their forests and using the land for productive activities, so why shouldn't we do the same?" the son of a large-scale farmer asked me rhetorically at a party in Belém. "There are many ways in which carbon can be captured, and insisting that the forest must stand is equivalent to threatening our nation," he repeatedly told me soon after we had met, in an effort to convince me that some of the NGOs I was observing were obstructing Brazil's progress.

Conservationism, on the other hand, was represented by some of the NGO personnel I spent my time with as a threat to the region as well—although not to such a large extent as developmentalism. The ethos of conservationists, to preserve large areas of forested land where no humans would be allowed to live, was considered by Nathan and many other NGO personnel as disconnected with the region's history and characteristics. And yet, some organizations still promoted it: "Over its history, Brazil has lost 95 percent of the Atlantic Forest (*Mata Atlantica*) [a large forested area on the eastern coast of the country], but we are attempting to create corridors for wildlife among private owners who by law must keep parts of their property as standing forest," said a speaker in a meeting of NGOs funded by United States Agency for International Development (USAID) in the city of Manaus. The owners he was referring to were mostly large-scale farmers. His organization thus advised these producers not only to comply with the law but also to ensure green corridors. These areas, however, were envisioned by conservationists as pristine forest without any human population.

Anthropological analyses have thoroughly criticized both developmentalism and conservationism as short-sighted perspectives (Orlove and Brush 1996; Hobart 1993). The development apparatus has been criticized as a one-sided effort that often disregards local knowledge (Hobart 1993) and generates narratives of success around large-scale projects (Mosse 2005).

Conservation without people has also been criticized for its lack of engagement with local populations (Chapin 2004; Wilkie et al. 2006). These and other critiques have helped shape adjustments to international agencies', governments', and foundations' policies in both arenas. The idea arose of an ecosystemic approach where local populations would manage resources in order to safeguard them (Ostrom 1999). From this perspective, biodiversity conservation itself is viewed as necessarily including local knowledge (Orlove and Brush 1996). There is, however, always the risk of romanticizing an image of the "ecologically noble savage" (Redford 1991) in local indigenous communities. This image is grounded in an assumption that indigenous people are on the one hand "good savages"—with innate goodness but nevertheless uncivilized—and on the other hand live in balance with their environment. While some cases demonstrate this assumption to be wrong, numerous organizations of indigenous communities have sought to use their traditional knowledge for a protection of territory and of their own social customs. This is part of what the socioenvironmental perspective is about.

Socioenvironmentalism differs from developmentalism and conservationism in a fundamental way: it denotes an aspiration for equilibrium between social development and the protection of the environment. It has its roots in the seringueiro (rubber tapper) movement that was pivotal in the formation of the "peoples of the forest" alliance against encroaching farmers and loggers during the 1980s (Schmink and Wood 1992, 103). Chico Mendes, this movement's most visible leader, became an icon that helped turn environmental concerns into demands of justice by indigenous groups.[1] Along the way, numerous scholars participated in the alliance, including a few anthropologists (Keck 1995; Allegretti 2002). Of particular value for the movement was the synergy that was generated among many different groupings that collaborated, adjusting narratives of struggles for indigenous rights, for example, to include environmental concerns. For Keck and Sikkink, this combination was the result of advocacy networks: "The recent coupling of indigenous rights and environmental issues is a good example of a strategic venue shift by indigenous activists, who found the environmental arena more receptive to their claims than human rights venues had been" (Keck and Sikkink 1998a, 18). Politically, socioenvironmentalism emerged out of a peculiar context in Brazil when, as the military dictatorship waned, there arose an aspiration for a democratic regime whose actors shaped numerous efforts to alter the institutional architecture of the government: "The new environmentalism

1. Many circumstances led to indigenous peoples' use of their environmental knowledge as part of their demands for land and improved circumstances. Posey's ethnobiological studies and his activism, for example, provided crucial analyses to understand the complexity of accumulated environmental knowledge among indigenous groups (2004).

emerged as the military dictatorship was ending, with an attendant rise in organizing initiatives within civil society in general, especially on the left. Second, the murder of Chico Mendes in Acre in 1988 generated widespread discussion of the links between the livelihood struggles of traditional forest peoples and protection of the Amazon. Third, the preparatory process for the United Nations Conference on Environment and Development (or "Earth Summit") in Rio de Janeiro in 1992 brought a wide range of environmental organizations, women's organizations, urban and rural trade unions, and other social movements together in sustained dialogue over almost two years" (Hochstetler and Keck 2007, 109–10).

The meeting in Canarana was laden with the region's history and current tensions. The city itself was born in the 1970s as a colonizing project organized by a southern Protestant pastor to cultivate the land in the region using a North American model of urbanization and land use.[2] In a few decades, the area surrounding the city changed from a thick forest to, mainly, fields that produced soy. In 1975, the first private "colonizing cooperative" in Brazil set out to bring southern farmers (*gauchos*) to populate an area in the Amazon in order to reduce tensions over land in the south of Brazil and to use what was considered pristine areas ripe for colonization. A Lutheran pastor and an economist organized the cooperative and the colonizing process. The hotel where I stayed during the meeting had been the first guesthouse in the area. In the lobby, they had postcards of a time when Canarana was only accessible by small planes. A photo showed several light aircraft parked outside the hotel. What was now a wide avenue outside the building used to be the runway. The meeting took place not far from the hotel in a large auditorium belonging to the Lutheran church, with a view of the square where a larger airplane that had originally brought many families from southern Brazil was now a monument to the foundation of the city. Not far from Canarana lies the Xingu National Park, the first area devoted to indigenous populations in Brazil (ISA 2011). It was created in 1961 after a campaign by the Villas-Bôas brothers to protect the area (Villas Bôas and Villas Bôas 1974; Garfield 2004). It is the largest national park for indigenous communities in Brazil, and it is home to sixteen ethnic groups.

During one of the breaks from the meeting in Canarana, an ISA worker I had met shortly before, João, explained to me how ISA helped local leaders within the Xingu Indigenous Park to create their own NGO, the Xingu Indigenous Land Association (ATIX). This was part of an effort to provide them with the tools to negotiate directly for various issues that concerned them. João said, "We do not want them [the indigenous communities] to feel that

---

2. For an analysis of migration to the Amazon, see Lisansky (1990) and Moran (1993).

we do everything, even if it is for them." ISA was founded in 1994 by activists and advocates who had worked for fifteen years in two different organizations focused on indigenous rights. Both IMA and ISA were the result of a trend that started in the 1980s in which new civil society organizations were formed through the merging of various small organizations or of individuals with experience in the sector. It was then that the movement to defend indigenous rights was linked to environmental concerns, thus framing the nascent "socioenvironmental" movement. It was the synergy of the "peoples of the forest" network that provided momentum (Mendes 1991, 47–48; Santilli 2005). João, originally from the state of São Paulo, explained his view of the meeting, and it encapsulated the socioenvironmental ethos: "What is at stake here is the confrontation between two world views [*cosmovisões*], you know? The destruction that takes place every day is heart breaking. You should have seen their reaction when we went to the site of a dam construction. They were crying, the floor fell from under their feet. They could not believe what their eyes were seeing, it was hard for them to walk on that barren landscape. To see not only felled trees, but the earth removed and bulldozed, everything. . . . There was wailing and despair. I also cried. We took a small group there by bus, knowing that it would be painful, but it was difficult for all of us. Some of them wanted to go and attack the workers' camp, to do something to stop that . . . so they are here with a lot of reluctance; they distrust everything that is said."

And yet, there was hope. During one of the panels at the Canarana event, one indigenous leader was clear about his reproaches against prejudice: "We indigenous peoples always waited for you [white people] to call us to start a dialogue, but as indigenous people we are considered a minority, as if we were animals. We are human beings, and need respect," he insisted. The fact that so many indigenous peoples attended the launch of the Y'Ikatu Xingu campaign meant that despite the wariness, there was an underlying desire to collaborate. When I visited ISA's headquarters in São Paulo, one of its leading *indigenistas* (experts on indigenous issues) told me that the trust indigenous communities placed in them was the fruit of decades of collaboration. ISA is an organization that materialized from the merger between two previous groups dedicated to defending indigenous rights. One of them was the brainchild of the Villas-Bôas brothers, who had earned the respect and estimation of the communities with whom they worked. The work done by these brothers was far-reaching: "During their thirty years on the Xingu, the Villas-Bôas brothers developed a novel attitude towards the Indians. Although poor linguists and with no anthropological training, they intuitively established a rapport with these peoples, treated them as equals, learned about each individual's concerns, and made limitless time to converse with them" (Hemming

2005, 4). The socioenvironmental perspective therefore was the result of decades of experience by activists and advocates seeking ways to engage with concerns about the environment and people.

Socioenvironmentalism was part of a wave of political mobilizations during the 1980s and 1990s. It originally consisted of a repudiation of government policies and a demand for more comprehensive strategies that would allow forest dwellers to live without the threat of encroaching deforestation by loggers and farmers. This was framed in a general opposition to the military dictatorship that had ruled Brazil since the 1960s and that had set in motion the colonization of the Amazon in the 1970s. The Workers' Party (Partido dos Trabalhadores, PT) was founded in 1980 out of many clusters of workers' associations that sought Brazil's democratization. The union of seringueiros was part of this movement, and Mendes moved up the ladder to lead the general union of rural workers of the Amazon. In 1984, when the dictatorship ended and the transition to democracy started, the PT and many independent organizations that had worked together against the dictatorship focused on promoting a new constitution. The result was a national charter that is known for its support of citizens' organizations, which has allowed for a vernacular NGO blossoming.

When I visited Xapuri, the town in the state of Acre where Chico Mendes lived until the day he was murdered, among the local activists there was a mix of pride and disbelief—pride about what Mendes represented locally, nationally, and internationally; disbelief that the person who was accused of masterminding his murder, a local rancher in the area, was again free after years in jail and boasting about his business and the ongoing encroaching of the farming and soy frontier. Mendes's house is now part of a museum where visitors can learn about his life and political struggle. One can also see where he was murdered and read interviews with those who were closest to him. I went there with Fernão, an employee of the local branch of the World Wildlife Fund for Nature Brazil (WWF), who was keen to show me the history. "There was an important chapter of the socioenvironmental movement written here, and sometimes it seems that we have not yet learned the lesson," he said. He also introduced me to Antonio Alves, a local intellectual who was one of the guiding lights of the movement. As a poet, journalist, and activist, Alves sought to put into words the socioenvironmental ethos that Mendes represented. One example is the concept of *florestania*, which he coined at the end of the 1990s. *Floresta* in Portuguese means forest. *Florestania* thus adapts the concept of "citizenship" (*cidadania*) to the situation in the Amazon because of its particular circumstances. In one of Alves's essays, he remarks: "In the end, what is that '*florestania*'? 'It is citizenship in the forest'—is the usual simple and quick reply. It's that, but it is also more. Apart from a collection of social relationships, rights, duties, laws, and conquests, *floresta-nia* is a feeling that can be expressed in the following manner: the forest does not belong to us, we are the ones who belong to her. That feeling induces us to establish not only a new social pact, but a new natural pact based on the equilibrium of our actions and relations in the environment we inhabit. It is a guiding feeling for our economic, political, and social choices—that is why I included citizenship—but also oriented towards our environmental and cultural choices—that is why it transcends citizenship" (Alves 2004, 129–30, my translation).

The basic principle behind socioenvironmentalism is that humans who rely on the forest for their subsistence are part of its ecosystem. In such a model, it follows that their lifestyle deserves to be protected along with the forest they inhabit while large, external corporations' rights are not given the same considerations (Santilli 2005). This principle applies not only to the indigenous communities spread throughout the Amazon forest but also to forest dwellers brought in by other circumstances, such as rubber tappers. In Manaus, Fernão introduced me to a friend of his who lived in Rio de Janeiro and who owned a business. It was a social enterprise that manufactured goods from rubber extracted in the traditional seringueiro way from the forest, and it was sufficiently successful that a local community in Acre could make a living from it. Fernão's friend showed me the catalog of fashionable objects that were made from the soft, leather-like fabric, such as bags, belts, and even cell phone protectors.

This type of partnership between communities seeking a sustainable form of sustenance and businesses seeking to support them is often mediated by NGOs. One of IMA's projects, for example, is to help a few families in a protected forested area in the state of Pará to use the precious wood of naturally fallen trees to make furniture that IMA then sells, either on the internet or in partnering stores in large cities in southern Brazil. On one occasion, while on a week-long boat trip down the Tapajós River to visit various IMA projects, the boat I was on stopped to pick up a few furniture pieces. "These are treasures they make," Alina said, while watching the locals transport stools and chairs to the boat. Every house and apartment of IMA's staff that I visited had several of these pieces of furniture. IMA undertakes studies to certify the project's commitment to sustainability and uses their findings to market the stools, chairs, and tables as ethical goods. Those buying them thus know that they are supporting forest dweller families in the Amazon. These partnerships, which have become common and are now known as "ethical consumption," are usually certified by such groups as Fair Trade (Bird and Hughes 1997; Hartlieb and Jones 2009; Harrison, Newholm, and Shaw 2005).

In the 1980s, however, collaboration between different groups in the

Amazon was just starting to take place. The significant step that helped cat-
apult Chico Mendes to political stardom was the coalition he formed by in-
volving indigenous groups, *quilombos* (communities of former slaves who
had managed to escape), and small farmers—alongside the rubber tappers—
in the protection of the rain forest and the traditional ways of life it hosted.
The alliance became known as the "peoples of the forest" (Mendes 1991, 47–
48). One of the movement's flagship demands was the creation of "extractive
reserves," or protected areas where local communities would be allowed to
maintain sustainable activities that would not harm the forest ecosystem
(Mendes 1991, 43; Hemming 1985b). Since Mendes's assassination on December 22, 1988, the Brazilian government has sanctioned several of these
extractive reserves (Brown and Rosendo 2000; Hall 2007; Vadjunec and Rocheleau 2009). These reserves have been at the center of the socioenvironmental movement as examples of the practice of sustainability.

As Fernão and I drove back from Xapuri to Rio Branco, I asked him about
the lonely trees I saw in deforested areas where cattle were grazing. "Those
are Brazil nut trees. Did you know they are a symbol of sustainability? Listen,
man, those there were left standing because Brazil nuts are so valuable that
they are considered a good source of income. But they are very social trees; if
they are not surrounded by standing forest they will not produce nuts. That
is why when you buy Brazil nuts, you know you are helping to keep a forest
standing." It seemed to be a slogan for merchandising Brazil nuts. Sadly, we
saw it was too late for the remaining lonely trees, as owners of that land consider livestock more profitable than Brazil nuts. Cattle ranching is one of the
largest problems in the Amazon. While doing my fieldwork in Brazil, news
came out that there were more heads of cattle than people in Brazil. As a con-
sequence, the methane from cattle was the second highest source of pollution
in the country, just after the carbon dioxide from the burning forest.

When Luiz Inácio "Lula" da Silva took office in 2002, he appointed a minister for the environment who had a socioenvironmental agenda. It was Marina Silva, who had grown up in the forest in a seringueiro family and had
been an activist who was close to Chico Mendes. She was the youngest-ever
elected senator in Brasília. Her work in the ministry was geared toward bringing the government in line with socioenvironmental principles. In several
meetings I attended, I met one of her close confidants from the ministry, who
worked as a special envoy to the Amazon region. She also used to be an environmental activist and sought to collaborate with NGOs such as IMA to stop
deforestation and improve forest dwellers' well-being. "As activists, we knew
that in order to change policies we may have to join the government, and this
is where we are now," she told me. Many environmental activists and advocates joined Marina Silva in the ministry. They drew up numerous strategies

for improving the government's approach to socioenvironmentalism, for example: improved system of protected areas (Silva 2005), policing to avoid deforestation, and stricter license regulation for farming and other activities. The rate of deforestation was reduced, but not stopped. In May 2008, Marina Silva quit Lula's government in protest because other government ministries undermined her demands for a "transversal" strategy to promote a socioenvironmental plan of action for Brazil. The income generated by soy production is but one reason for the reluctance of state government officials to put a halt to its developmentalist policies. On August 19, 2008, Marina Silva quit the Workers' Party and decided to run in the presidential election in 2010 under the Green Party. She obtained 19.4 percent of the votes. In 2013, she sought to launch a new political party that she called Rede Sustentabilidade (Sustainability Network), a party name that points to a growing awareness of networks in Brazil as fundamental for socioenvironmental purposes. However, this campaign failed because she did not receive enough signatures of support. She ran again for president in 2014, this time under the Brazilian Socialist Party, but ended up in third place with 21 percent of the votes.

Even though it is a popular perspective in Brazil, socioenvironmentalism is still not part of the nation's mainstream policies. "It has been a struggle, but we have advanced considerably," Fernão told me. Among those promoting the socioenvironmental perspective, Chico Mendes and his work remain the most potent symbol. Chico Mendes won international acclaim when he addressed the Inter-American Development Bank in March 1987 in an attempt to convince its representatives to consider the preservation of the forest and the livelihoods of the local population before they approved a road project (Souza 1990). He was brought to the United States by an American anthropologist who is now a member of IMA's council and is active in other environmental NGOs in the United States and the Amazon. Mendes's personal story reflects the reality of the circumstances prevalent in the Amazon region in the mid to late twentieth century. He only learned to read at the age of twenty, from a very opinionated seringueiro during a time of political unrest against the military dictatorship. Later in life he organized a union of rubber tappers and struggled to get education programs running to encourage members of the community to become aware of the injustices that characterized their situation and to seek better living conditions. He became an influential leader as he learned to work in a decentralized way in order to avoid the movement being decapitated.

The movement that Mendes led relied on support from abroad for legitimacy, resources, and protection from hostilities by landowners and the government itself. A few days before being shot dead in his house in Xapuri, Mendes said in an interview: "I'm afraid we have had more support from

abroad than from people in Brazil. . . . It was only after international recognition and pressure that we started to get support from the rest of Brazil" (Mendes 1991, 51). Oxfam, for example, provided grants for the rubber tappers' education programs (Mendes 1991, 34), while Christian Aid
helped fund a cooperative run by the peoples of the forest coalition (Mendes
1991, 76). A few years before Mendes's death, Gro Brundtland—the former
prime minister of Norway—had interviewed Mendes during the hearings of
the World Commission on Environment and Development, for what is titled "Our Common Future" but is known as the Brundtland Report (WCED
1987). This report established the notion of "sustainable development" based
on projects such as the extractive reserves. Chico Mendes's influence thus
lived on after being assassinated on December 22, 1988, by a local rancher
who felt threatened by his activism. His murder led to international outrage
and pushed the government to give in to some of the rubber tappers' demands (Andrade et al. 1989).

Violence in the Amazon region has long been considered an element of
its frontier zone status (Hemming 1985a, 1987; Schmink and Wood 1992).
People living in isolated towns thousands of kilometers from any city told me
repeatedly about the parallels they see between their life in the Amazon and
their image of the American Wild West. The history of socioenvironmentalism is inextricably linked to the area's past and the manners in which it is
interpreted. As a frontier zone, it is often analyzed through a political economy lens (Foweraker 1974, 1981). Government policies to populate the area
in the 1960s and 1970s combined fears over a possible foreign invasion with
an interest in spurring economic activities. Thousands of families thus relo-
cated to the area with the dream of improving their lives by working the land.
Perhaps the fear about the potential loss of the Amazon territories to foreign
powers that fueled the military's colonization projects was due to the region's
history. In the early nineteenth century, a rebellion took place known as the
cabanagem (Harris 2010). Although the movement was not a separatist one,
its force at a time of uncertainty after Brazil's independence from Portugal in
1822 sparked fear about its potential loss. In 1835, rebels achieved the capture of Belém and large swaths of land, but only for approximately half a year
(Harris 2010, 15). What remained from that experience was an apparent resolution by landowners and government officials to be ruthless in controlling
the population. When the rubber boom followed, violent subjugation was a
clear strategy for keeping workers at bay.

Historical accounts have shown different interpretations of what has happened in the region. Schmink and Wood's (1992) revision of the history of the
Amazon focused on invasions for resource extraction. Throughout their volume, they discuss how at different stages during the last few hundred years,

indigenous communities and migrant workers have been constant victims of grander development projects of military regimes. Anderson (1999) referred to the effects of colonization from the eighteenth to the twentieth centuries as an exploitation of the area. Other accounts analyze the mix of interests and dynamics that have converged in the struggles between developers and defenders of the forest (Hecht and Cockburn 1989). Cleary (1990), for example, offered a description of the Amazonian gold rush of the late 1970s. For Furley (1994), the isolation and settlements in Roraima, the northernmost state in the Brazilian Amazon, exemplify the mix of developments affecting the whole region. The literature is vast, and it is not the purpose here to offer a review of academic approaches to understanding invasions of the Amazon. It is relevant, however, to mention a few key studies that have recorded the political and economic forces that have crisscrossed the region, leaving their mark on its natural and social environments. Hecht and Cockburn (1989) classified all those who intervened in the area as either developers, destroyers, or defenders of the Amazon. Price (1989) recounted his experience as an anthropologist trying to defend a small society from a development project backed by the World Bank. Meggers (1996) followed up on a book originally published in 1971; calling the region a counterfeit paradise, Meggers argued that Amazonian social groups have adapted to a changing world through cultural innovation. With respect to development projects, Bunker (1988) argued that they represented the failure of the modern state. Anthropologists have been frequent witnesses to regional changes. After Pace (1998) visited a town previously studied by Wagley (1958), he noted many of the changes that have occurred over the last few decades. As local inhabitants sought to improve their living conditions, some turned to cooperatives, others to liaising directly with markets, and yet others sought to stop large-scale exploitation of lands they considered their own.

Cleary (1993) pointed out a set of problems raised by a previous political economy analysis of the Brazilian Amazon, disapproving of the tendency to simplify *class* and *peasantry* in a region where both are somewhat fluid terms (Cleary 1993, 336). Cleary also argued the monetarization of the region did not necessarily imply the infiltration of capitalism, as it was mostly the informal economy that fed economic growth at the time of his article (343–45). If there were any doubts back in the 1990s, however, the recent rise of soy production has meant a tidal wave of market capitalism in the region. Some proposed solutions, such as carbon emissions trading or the payment by the high-emitting nations to others that protect ecosystems like the Amazon forest—are also part of this process.

The key event that turned the Amazon rain forest into a sought-after territory was the rubber boom in the nineteenth century (Santos 1980). Its legacy is visible throughout the cities of Belém and Manaus, where grand buildings and houses in a French style dot the old centers around large opera theaters. IMA's headquarters was one of the houses built by rubber barons. In the nineteenth century, the Amazon forest was the only region in the world that could produce enough rubber for the European industrial revolution. In 1827, thirty-one tons were exported from Manaus and Belém. The amount increased to 156 tons in 1830; 388 tons in 1840; 1,446 tons in 1850; and it reached 2,673 tons in 1860 (Weinstein 1983, 9). After that, amounts skyrocketed, reaching 10,000 tons by 1879. The foreign demand for rubber took place long before automobiles became popular (Hecht and Cockburn 1989). Indeed, the first mass-produced item to use tires requiring rubber was the bicycle: "In 1894 there were 250,000 bicycles in France; twenty years later, there were almost five million" (Hemming 1987, 273). The rapidly increasing demand was a challenge for producers because trees did not yield rubber in plantations, so it had to be extracted within the forest in a similar fashion to that which I described for the Brazil nuts. Someone (a rubber tapper) would walk through the forest, visiting a series of trees from which he or she had made markings to allow for rubber to flow out and into a container. The rubber tapper thus collected the rubber and prepared it for transportation. This extraction method did not cause damage to the forest, but it did bring in immigration flows that continually confronted indigenous populations (Eden 1990). An international drop in price that virtually ended the Amazon rubber boom was caused by the first documented case of biopiracy: in 1876, Sir Henry Wickham stole approximately 70,000 rubber tree seeds from the Santarém area and brought them to the Royal Botanical Gardens at Kew (Dean 1987; Jackson 2008). There, they were altered to be productive in plantations and were sent to Sri Lanka, Malaysia, and other tropical destinations controlled by the British Empire.

Rubber extraction in the Amazon relied on cheap labor that was enforced through violence and accompanied by miserable living conditions (Davis 1997). At first, indigenous peoples were forced to carry out the extractions. Although numerous individuals managed to escape, it is estimated that enforcers killed tens of thousands. Consequently, labor shortages began to affect the extraction of rubber. It was in response to this shortage that new waves of immigrant workers were brought to the Amazon region, who have since then been known as seringueiros (rubber tappers). There was a new rubber boom during World War II, but it only lasted a few years. This boom was due to Japan's invasion of Burma (Myanmar), where it managed to briefly control the rubber plantations.

During most of the twentieth century, Brazil went through political upheavals with several military coups d'état trumping elected governments. In

1961, the government, led by the head of the Labour Party, João Goulart, tried to undertake structural economic and social reforms as well as to help workers' organizations challenge their oppressive situations (Mendes 1991, 20). During that time, innovative community action and literacy projects sprouted everywhere, led by progressive intellectuals like Paulo Freire. In response, conservative forces and the military carried out a coup d'état in April 1964 to stop what it considered a left-wing "swing toward a radical nationalist strategy" (Skidmore 1988, 17). The dictatorship mixed a populist tone (through the public aspiration of a redistribution of land) with a nationalist one (used to legitimize its colonization program in the Amazon forest). The redistribution of land only partially happened, and it did not include a systematic allocation of legal proof of land tenure. Unsurprisingly, juridical problems continue to this day, with many competing claims for the same plots of land and new arrivals seeking to make a living from these properties.

From 1964 to 1984 (the years comprising the country's most recent dictatorship), the military attracted foreign capital for rapid industrialization while maintaining political stability and cheap labor. Not surprisingly, this was accompanied by a military structure that clamped down on dissent—in particular on Maoist groups hidden in southern Pará (Bourne 1978). In this climate, the Catholic Church provided support for groups that opposed the "business as usual" model that prevailed in the government. Many priests and bishops denounced human rights abuses and demanded social justice around the subcontinent. In 1968, the Latin American Episcopal Conference held in Medellín, Colombia, pushed for a progressive interpretation of the Second Vatican Council. After this event, the creation of Comunidades Eclesiais de Base, or Christian Base Communities (CEBs), skyrocketed (Encarnación 2003). By the early 1980s, the number of CEBs in Brazil was estimated at around 80,000 (Burdick 1993). These organizations provided a training ground for leaders of what would become Brazil's most influential social movements and civil society organizations. I have called it elsewhere the training for a networked capacity infrastructure (Acosta 2013). It was from a mix of these movements and leftist intellectuals that the Workers' Party or Partido dos Trabalhadores (PT) was born in 1980.

In this milieu, socioenvironmentalism became a frontline for a broader struggle for new rights. As has been long acknowledged: "A striking phenomenon amongst Brazilian NGOs in the late 1980s, as their role has expanded to encompass wider political and policy-formulation tasks, is the growth of the environmental lobby, embracing not just ecological but also related social and indigenous questions" (Goodman and Hall 1990, 13). The associational networks that first pushed for the end of the dictatorship and then the new constitution set an agenda of juridical reform. As Santilli notes, "The influence of socioenvironmentalism is felt in the [Brazilian] Constitution, which established a solid basis for the recognition of socioenvironmental rights and for the systematic interpretation of environmental, social and cultural rights, in the infraconstitutional legislation that gave them more precision and efficacy" (Santilli 2005, 19, my translation). Socioenvironmentalism is thus placed within a framework of pluralism and multiculturalism where awareness of traditional knowledge and territories is based on anthropological and biological studies. It seeks forms of "good governance" in the management of environmental reserves (Santilli 2005, 35). Its advocates consider it an emerging paradigm of "ecosocialist development" as opposed to the apparently prevailing "capitalist expansionist" one (Santos 2000). It may well be regarded as a form of "transenvironmentalism," or a struggle that through environmental issues engages with other spheres of concern such as democracy and rights (Kousis and Eder 2001; Voulvouli 2009).

The mix of traditional and scientific knowledge formed the backbone of the socioenvironmentalist movement. Indeed, academics have been strongly involved in the skyrocketing of NGOs and other civil society organizations over the last three decades. As well, thousands of organizations are dedicated to the promotion of human rights, urban issues, indigenous rights, gender issues, and a range of other topics. In many of these groups, anthropologists have been strongly engaged. The socioenvironmental movement gained momentum when Brazil hosted the 1992 UN Conference on Environment and Development, known as the Rio Summit (Little 1995). The meeting served as an international reminder of the relevance of the Amazon to the world. A direct repercussion in Brazil was an influx of international funds to help study and protect the rain forest. The incoming money was partly dedicated to starting local NGOs. Founded in 1994, IMA and ISA were part of the wave of regional NGOs that started after the conference. IMA's stated mission implied a socioenvironmental take: "To contribute to the development process in Amazonia, by taking into account the population's social and economic aspirations, as well as maintaining the functional integrity of the regional ecosystem by means of research, outreach and education."

As an NGO dedicated to carrying out its own scientific research for advocacy purposes, IMA relies on the legitimacy its publications provide (see chapter 4 herein). With it, IMA bridges global concerns with local interests. There is no doubt that the current situation of the Amazon, the largest standing rain forest in the world, is part of a geopolitical struggle, with strong elements of economic bonanza—or a financial windfall—stemming from natural resource extraction (e.g., mining and logging) and large-scale farming as well as from an increased worldwide interest in climate change that puts pressure on the Brazilian government to stop deforestation in the region (Barbosa

2000). For example, the fate of rubber has led to caution over biopiracy. Scientists need special permits to take samples or animals. Also, recent regulations aim to clamp down on companies that use traditional plants without paying royalties to indigenous communities whose knowledge led to their discovery. This is part of a global trend to acknowledge by law, and pay royalties for, traditional knowledge (Kamau and Winter 2009). It is one of the issues that was negotiated for the Convention on Biological Diversity, which was signed by 150 government leaders in 1992 as part of the activities of the Rio Summit (CBD 1992). Despite these efforts, and although there have been some cases of fines for companies and preventive actions, it is hard to know the effectiveness of such measures.

All this international attention on the Amazon has raised the expectations for sustainable regional development (Hall 2000). Nevertheless, the soy frontier advances and industrial production in Manaus keeps growing. These trends are combined with an urbanization that increases risks for the Amazon biome (Browder and Godfrey 1997). The socioenvironmental agenda is therefore more of an ideal route by environmentalists and scientists than a government agenda. In having been incorporated as part of the government's discourse, it has lost its alternative edge. For this reason, several NGOs have made it their mission to keep insisting on its principles and implications. IMA is one of them.

The way in which the needs of indigenous peoples have been incorporated into wider demands has been informed by a more comprehensive understanding by scientists and environmentalists of the value of their experiences. Contrary to the manner in which many policy makers refer to these experiences, they are not limited to a knowledge of medicinal plants. International foundations and agencies have paid attention to recent expert appraisals of the value of local forms of knowledge for the protection of ecosystems. This is in part informed by results that anthropologists and other scholars have found over the last few decades on the effects of human populations on the Amazon forest. Previous notions of the rain forest as pristine have been discredited through evidence pointing to urban human settlements before the arrival of Europeans (Balée 1989, 2006; Balée and Erickson 2006). Through historical ecology, researchers have gained insights into long-term stages of activities in the forest through which various populations have played an active role in the making of the natural environment (Rival 2006). It is thus a case of dwelling contributing to the way an environment is shaped (Ingold 2000b). This must be considered as a central part of the socioenvironmental perspective.

In this chapter, I have explored the socioenvironmentalist movement from the experiences of some of those who live it. IMA and ISA are two examples of NGOs that combine the production of scientific knowledge with a social justice agenda to seek an improved balance between forest ecosystems and their human dwellers. The transformation of the Amazon forest over the last few decades has been a tale of destruction, with thousands of fires and ongoing deforestation. But in building a dense set of advocacy networks, activists and advocates have made sure to establish a basis for an improved understanding of the implications of such destruction. In many analyses by IMA scientists, some of them published by top-tier scientific journals, two sets of models appear: one named "business as usual" and the other one "governance." With a careful use of language to describe what is widely considered an unsustainable path, these scientists do not mask the bluntness of their warnings. And yet, today the Brazilian Amazon is under a more serious threat than had been imagined at the time of fieldwork. If there is hope for the accumulated knowledge about potential forms of regenerating its ecosystems, it might just be in the form of advocacy networks. In their plurality and complexity, these organizational arrangements strengthen campaigns and shed light on specific problems; they also multiply a series of personal experiences in alternative political performances that may be crucial to redress some of the damage done to the area.

2

Barcelona and Its Nationalist yet Progressive Milieu

I want to point out that messages in our network need to be written in better Catalan. As they are, it will not do if we want to convince [government] authorities to help us financially.

—María, NGO advocate and member of the FSMed Technical Secretariat

ON JUNE 15, 2005, the day before the official start of the four-day Mediterranean Social Forum (Fòrum Social Mediterrani, FSMed) event in Barcelona, its Spanish organizers and the delegates from abroad were invited to attend an official ceremony at the Palau de la Generalitat de Catalunya (the headquarters of the regional Catalan government). I walked there with a group of union leaders, feminist activists, and others who had been working over the previous years to make the FSMed happen. Just as we were about to go in, one member of the group, a Palestinian activist, announced his departure: "This type of official ceremony is what we are against." Some in the group smiled back at him, making gestures of understanding as they said farewell. "See you tomorrow," we said. The rest of us went in, and we were led to a grand room where a few minutes passed before a high-ranking Catalan government official gave a speech. It was a solemn affair. Around half of those present at the ceremony were new faces; I had never seen them during the fourteen months I had been following the preparations for the FSMed. Many were government officials. The contrast in dress code was clear: while those I had not seen before wore suits or formal attire, the FSMed's organizers were attired in a wide variety of garments ranging from slashed jeans and black T-shirts by anarchists, to casual wear by NGO workers and activists, to semiformal garments by support personnel.

The official speech was about Catalonia's "open arms to the Mediterranean" and about Barcelona's key economic, cultural, and political role for the region. The tone of the speech seemed to indicate that the FSMed was actually part of the general strategy of the Euro-Mediterranean Partnership,[1] something that surprised more than one of those present from the organizing committee. In many meetings during the previous year, I had heard numerous criticisms about such a partnership. In fact, the FSMed was often portrayed as a group that had officially denounced such a partnership. At the reception held afterward in the halls of the palace, I met a member of the Institut Europeu de la Mediterrània (European Institute of the Mediterranean), a Catalan government-sponsored think tank that specialized in Mediterranean issues (IEMed 2013). After a bit of small talk, he seemed eager to share with me, a researcher, the underlying purpose of the FSMed: "We will show them all what Catalonia is capable of," he told me without clarifying either who "them" was or what he meant. However, from context, I gather that he meant that "we will show the international community what Catalonia, as a leader in the Mediterranean, is capable of." Over the course of my fieldwork in Barcelona, I attended several meetings and public events in which speakers would refer to the city as the most important port in the Mediterranean. It was common in talks about trade, diplomacy, and tourism, but also in discussions about humanitarianism and solidarity movements. In the ceremony just mentioned, local government officials appeared to signal that as a region that had faced its own struggles, Catalonia had a special solidarity and sympathy with different struggles around the Mediterranean. That is why Catalonia had invested tens of thousands of euros to ensure the FSMed would take place in Barcelona.

This chapter lays out the setting in which the four-day FSMed event took place. In it, I interweave the activist and advocacy milieu in Barcelona I witnessed while helping organize the FSMed event with the history of Catalonia's progressive movements and its uses by its political class. I argue that several grassroots initiatives were incorporated into nationalist narratives to advance the cause of Catalan independence. The sense of widespread solidarity that Barcelona and Catalonia are historically known for has not only literally shaped the region in its infrastructures and services but also in the narratives that its leaders use to portray its national character. One example is its take on urban design, which became known as the "Barcelona prototype" or "modelo Barcelona" (Monclús 2003; Montaner 2004b; Capel 2005). From the fall of the Franco regime, several urban planners incorporated the opinions

1. The Euro-Mediterranean Partnership was the official name of what was known as the Barcelona Process because it was started in that city in 1995. It consisted of an agreement to strengthen relations between the European Union and members of the Mashriq and Maghreb regions. It was relaunched in 2008 as the Union for the Mediterranean (UfM 2013).

and needs of urban dwellers to their redesigns of avenues and public spaces. They did so by paying attention to the voices of the well-organized neighborhood associations (Alabart 1981). This style of urban design gained a global reputation when Barcelona hosted the Olympic games in 1992. It was also the first time a city used the considerable investment from the games to reorganize urban infrastructure and services. When I started my fieldwork, in 2004, preparations were in place for a new type of massive event through which city authorities hoped again to use investment for urban regeneration. This event, however, was to be something completely new, drawing from international fairs. It was called the Universal Forum of Cultures, or Forum Barcelona 2004. It was sold as an event highlighting cultural diversity and solidarity, thus somewhat mimicking the World Social Forum. For the radical activists who were involved in the FSMed process, it was a blatant effort to merchandise the progressive spirit of Barcelona. For many other critics, 2004 was also the year when nationalist and capitalist interests overtook those of urban dwellers in thinking about the city (Andreu 2004b). This context illustrates my analysis about the use of the FSMed by politicians and some activists to position Catalonia as a leading supporter of civil society organizations from around the Mediterranean. The relation between Catalan nationalism and civil society has been pointed out before (Pollock 2001; Guibernau 2012). As we will see later on, the effects of this relation for advocacy networks provides examples of frictions and tensions.

A Leading Progressive City

As the capital of Catalonia, Barcelona is also the most visible international outpost of Catalan identity. During my fieldwork, the city's political class appeared eager to promote it as a global city with a number of positive features: major cultural attractions, in the form of museums or buildings; expertise in various creative industries, such as industrial and web design; and a strong commercial potential. The activists and advocates who joined efforts to make the FSMed happen in Barcelona came from a wide spectrum of ideological camps and organizations. Their viewpoints were as varied as the background from which they came—although, as I have mentioned previously, they often coalesced in two clusters around radical socialist and pragmatist social democrat agendas. For the former, the European Union's efforts to establish a collaborative network with countries of the Maghreb and the Mashriq were blatant efforts to tap into new markets and reach deals that would benefit the EU, such as that of producing solar power (MSP-PPI 2013). Activists in this cluster tended to distrust EU-funded projects for those regions around the Mediterranean that were outside the European Union. For the social democrat advocates, the promotion of political rights and freedom go hand in hand with an interest in markets. Fair trade, for example, was promoted by these groups as a form of community empowerment.

The many groups participating in the FSMed's Technical Secretariat—the forum's de facto organizing committee that was based in Catalonia—were also involved in many other efforts or projects, either of their own or of different networks. They were among Barcelona's leading activist and advocacy groups that were in some way or other interested or involved in Mediterranean issues. The city of Barcelona is known among activists for being a hotbed of activism (Juris 2008a, 2). It is the only city in the world to have had a municipal government led by anarchists (which may be considered a contradiction in terms). There is pride among the city's activists about the city's history of dissent. In 2004, for example, one group of alternative artists set up a tour called the "route of anarchism" in the city (Obrador and Carter 2010). Many of the places highlighted on this tour had been sites of violent confrontations. Such outspoken expressions of appreciation of subversive actions, however, have only become possible in recent decades. During the regime led by Francisco Franco from 1939 to 1975, there was a strict control of dissenting views through violent repression and censorship. After Franco's death in 1975, Spain went through a transition to democracy. Barcelona's streets, like others around the country, were suddenly full of an invigorated sense of freedom. In 1977, for example, there was a libertarian conference held in the city's celebrated Park Güell. Víctor, an activist who was one of my main interlocutors, had taken part in that gathering. He particularly remembered what were called *jornadas libertarias* (libertarian days), which were a series of talks and informal workshops that Noam Chomsky attended. "We lived in an atmosphere not of anguish, but of 'we'll have one, this one.' And it was a big party in Park Güell, four days of total havoc: drugs, sex, everything. . . . The Ramblas went mad from 7 pm. It was great, lots of fun, a blast, and a *joie de vivre*, of sexual liberty and promiscuity." For Víctor, this came to an end because of a government strategy to put down the revolutionary fantasies of the youth culture: "The movement had a strong potential. It was thus very dangerous to the system. There was a need to control it, and it was not easy. In 1977, the Spanish political system lost control of Barcelona, in the sense that this city believed that liberty was possible under the capitalist system, or so we believed. Turned out it's not." He claims that what dampened this revolutionary atmosphere was the systematic distribution of heroin, which hooked a wide margin of those involved in the movement. "For two months, doses of heroin were distributed for free in the Ramblas. Luckily I did not get hooked, but all, well, a big part of the avant-garde of the movement, the most vital membership, stayed hooked on heroin."

Despite the reduction of activity during the following years, what persisted was an intense sense of solidarity and activism. During the 1980s, Barcelona was shaped by the democratizing process, the fact that Spain joined the European Union, and by local activists. Many of the activists involved in the FSMed organizing process either had a long experience in activism or were a part of families where activism was important. The sense of solidarity was embedded in a cosmopolitan awareness of the needs of less privileged regions outside of the European Union. Several activists I met in the milieu surrounding the FSMed had experience traveling to small communities in Latin America, Asia, or Africa to assist on projects or establish working relations with local groups. Most of the local associations involved in the FSMed had projects or alliances with groups from the Maghreb, the Mashriq, or the Balkans.

Solidarity was also a guiding principle among existing collectives within Barcelona or Catalonia. As months went by in which I attended meetings, rallies, alternative venues, or NGO workshops, it was evident to me that many groupings coexisted in complex networks of collaboration and independent projects. I call this an "advocacy ecosystem" in Barcelona as a way of addressing what appeared to be a unique setting (physical environment) where interactions took place between a variety of activist and advocacy groups and individuals (organisms). Rather than indicating an order or design, *system* refers here to a complex web of interactions in which different actors are involved, to a variety of extents. In this sense, the concepts of network or meshwork did not address the variety of sizes among groups involved and the complexity of their interactions. In recent decades, ecological studies have sought to understand different scales of interaction between organisms in their habitats (Petersen et al. 2003). In this sense, rather than implying a bounded system, what is sought by this metaphor is to refer to the various interactions that take place in a given setting. Barcelona's history of dissent and activism has allowed for the development of what I call *advocacy service providers*. These are businesses and organizations that work under a solidarity economy scheme, which entails providing services for social movements or NGOs at a much lower rate than for normal clients. One example, which I have written about in other chapters, was a printing service the FSMed secretariat used for the production of stickers, flyers, and even the meeting's program. As part of the team dedicated to such products, I went to the shop and saw numerous posters and flyers for other mobilizations, protests, or alternative events. The shop also had commercial products, for which they charged full rates. The fact that the shop was well known among activists and advocates showed its continuous support of local nonprofit activities. Other examples of similar support services were several cooperatives that provided catering for meetings, or even Babels (Boéri 2008, 2009), a translator network formed by professional translators who volunteer their time and expertise for events such as the World Social Forum (WSF).

The local activist milieu also received official support. The city government of Barcelona and the regional Catalonian government, for example, promoted opportunities for the region's youth to volunteer for a variety of independent organizations within Catalonia or abroad. Various youth offices had leaflets advertising these opportunities and provided information about where to get related advice. These offices also helped organizations apply for funds from the European Union for volunteer exchanges. During my time in Barcelona, I helped Víctor fill out several applications for EU funds to bring volunteers for Socis de la Terra (SdT). One of them was for a Palestinian to help out in clerical activities of the FSMed organization process for six months. Our application was successful, and a young university graduate from Nazareth was able to live in Barcelona to help out with translations into Arabic and contact various activists and advocates from the Arab world.

In order to access funds and support from local and regional governments, as well as from the European Union, civil society organizations have to conform to very specific guidelines. These regulations, which are intended to promote civil society organizations in a way that fits governmental bureaucratic standards, are interpreted by radical activists as controlling and an administrative burden. The rules through which EU institutions seek an orderly and accountable milieu of nonprofit activities appeal to the most established and institutionalized civil society groups (Armstrong 2002; Smismans 2003). For radical activists, these regulations are burdensome and generate distrust. For reformist activists, however, such rules are simply part of their work.

Numerous reformist groups in Catalonia are Catalan nationalists demanding issues such as the adoption of Catalan as an official language in the European Union. A constant remark I often heard in public debates or even discussions among secretariat members respecting this issue was that more people speak Catalan than Danish in the EU. Several members of the FSMed secretariat were involved in nationalist Catalan groups. One of the most active members, María, worked at the time of my fieldwork for the Escarré International Centre for Ethnic Minorities and Nations (Centre Internacional Escarré per las minories ètniques i les nacions, CIEMEN). On its webpage, this organization announces that it advocates for ethnic minorities and their languages the world over (Ciemen 2013). One of its main activities is its role in Mercator, a European network that defends minority languages such as Welsh and Frisian (Mercator 2013). Its webpage, however, is only in Catalan, and most of its activities are dedicated to furthering the Catalan identity. In so doing, it has clearly helped strengthen the case for Catalan nationalism (Conversi 1990). María's active participation in and commitment to the

FSMed appeared to have the purpose of giving the secretariat a distinctly Catalan character.

In the eyes of union members and other left-wing activists, María's role resonated too strongly with Catalan nationalist movements. These activists felt more strongly about the region's leftist traditions. The local labor movement is long-embedded in Catalonia because it was one of the regions in Europe that developed a strong industrial base early on in the industrial revolution. In 1869, Giuseppe Fanelli, an envoy of Mikhail Bakunin, helped organize the workers' movement in the region for the socialist First International (Reventós 1987). This leftist tradition made Barcelona the stronghold of republican resistance during the Spanish Civil War. This period and the many conflicts among the different leftist groups at the time were well portrayed by George Orwell in *Homage to Catalonia* (Orwell 1938), which was his personal account of experiences and observations of the Spanish Civil War. Orwell successfully portrayed the conflicts among left-leaning movements, as there was much mutual suspicion and hostility among them. A wide variety of leftist tendencies in the region—although not all—have links to nationalist agendas (Crameri 2000, 30).

Barcelona is also well known for its urban design and creative industries. Because of the affluence generated through its industrialization, city authorities held two universal expositions, or world fairs, in 1888 and 1929 (Grau and López 1988). Buildings from the turn of the twentieth century, including those of Antoni Gaudí, have now become symbols of the city or even UNESCO World Heritage sites. After Spain's Civil War, the Franco dictatorship favored a centralization of activities and resources in Madrid and castigated Catalonia and Barcelona for their fierce republicanism. Once the Franco regime fell, the socialist party coordinated efforts by a group of intellectuals to enhance the city's prospects for growth. This led to what has been known as the "Barcelona prototype" (*modelo Barcelona*) of urban development (Smith 2005; Borja 1995, 2003; Monclús 2003).

Those behind the "model" referred to the inclusion of citizens' participation as its raison d'être. For Pep Subirós, one of its creators, "Barcelona constitutes one of the most consistent and promising trials conducted to date of what could be a modern democratic city, namely, a project where form is as important as function, beauty as utility, the city as its citizens" (Subirós 1989, 106). It was a time when urban movements clustered in networks such as the influential Federation of Associations of Residents of Barcelona (Federació d'Associacions de Veïns i Veïnes de Barcelona, FAVB) (Alabart 1981). The dense cluster of networks that developed was a key ingredient of what was later called the "informational city" (Castells 1989). Policy makers around the world followed the "model" as it combined national or regional macroevents as tools for urban regeneration. The first of these macroevents was the Spanish FIFA World Cup in 1982, for which two Barcelona stadiums were used. A few years after, in 1992, Barcelona hosted the summer Olympic games while Seville held a World Fair (Harvey 1996). In between both events, in 1986, Spain had joined the European Union, allowing for an influx of funds and assistance that facilitated even further urban growth and redevelopment.

The government of Catalonia took advantage of its role as Olympic host to show its national identity to the world (Hargreaves 2000). An agreement between the city government, the Autonomous Government of Catalonia (Generalitat), the Spanish state, and the International Olympic Committee made allowances for the Catalan flag to be displayed and for the use of Catalan as one of the games' official languages, alongside English, French, and Spanish (Llobera 2004, 4). Barcelona's successful and profitable Olympic Games set a trend among cities and countries to compete against each other to host them (Del Olmo and Rendueles 2004). Up until then, the games had left cities with large deficits, as was the case with Montreal in 1976. This trend had resulted in Los Angeles being the sole candidate for the 1984 Olympics. Barcelona thus changed a trend and attracted attention from around the world. The *modelo Barcelona* was paraded around Latin America and even in the United Kingdom by Tony Blair to inspire international efforts for urban renewal (Andreu 2008).

The *modelo Barcelona* relied on the associative ethos that sprang from citizen groups into many areas of collective life (Marshall 2004; Puig Picart 2003). Such participation and involvement by city inhabitants in public affairs was but an extension of collective aspirations (Montaner 2004a, 2004b; Blakeley 2005). This was the milieu of activists and advocates involved in the FSMed organization process. All of them had, at one time or another, participated in massive demonstrations, protests, or consultations in Barcelona. By the time I was carrying out my fieldwork, however, the *modelo* appeared to be breaking down (Capel 2005). Among the activists I met, there was discontent about the way in which Barcelona had become commercialized. That year (2004), critical opinions began to be voiced ever more loudly. Intellectuals who had been behind the *modelo* no longer considered it viable. "Barcelona could achieve much due to mobilizations of its neighborhood associations and to a relative equilibrium of social classes. Today the city has changed and that model is no longer working," said Josep Ramoneda in 2004 (Pellicer 2004). As a journalist and philosopher, Ramoneda is an intellectual reference for Barcelona, as he was the director of a peculiar institution in the city: the Centre of Contemporary Culture of Barcelona (Centre de Cultura Contemporània de Barcelona) (CCCB 2013). This center is one of the city's institutions, as it houses innovative festivals, concerts, conferences, and debates, among many other activities. Its webpage claims it is dedicated to urban culture, but its activities include a wider range of topics than "urban culture" might suggest. During my fieldwork in 2004, for example, it hosted an exhibition called *At War*. Ramoneda explained in the catalog that the exhibition "seeks to show the [social] imaginary of twentieth century wars . . . the process of mental construction that leads a society to war and mobilizes its citizens in its wake, its cultural footprints in the experience and memory of societies" (Ramoneda 2004). Only a year before, in 2003, the city's inhabitants had come out in the hundreds of thousands—some even estimated one million protesters—as part of a global day of protest against the invasion of Iraq. In the same text, Ramoneda insisted the exhibition had been years in the planning and that it sought to avoid any position on Iraq. Its message, however, appeared to be critical of the destruction and implications of war. The first exhibition room, for example, included popular children's toys and games through history that had war as their theme, from lead soldiers to video games. Another room included a comparison between D-Day original footage and the initial sequence of *Saving Private Ryan* (Spielberg 1998), while yet another exhibited a multimedia testimonial section of women who had been raped in Africa. It also included a large section on Spain's Civil War and the scars it left on Barcelona.

The year 2004, therefore, was when the tide of opinion turned against the *modelo Barcelona*. Borja said that the promoters' interests had been overtaking those of the citizens since the end of the 1980s (Borja and Muxí 2004). In an interview years later, the urban innovation guru Toni Puig, who had worked for the city authorities and published enthusiastic texts on the *modelo* (Puig Picart 1992, 1995, 2003), said that 2004 was the year when the "decadence of Barcelona started" (Martínez and González 2013).

The reason for the turn of opinions was the celebration of one more macroevent in the city, which was particularly relevant to my research project: the Universal Forum of Cultures, or Forum Barcelona 2004 (ForumBCN 2013). The logo of the Forum was everywhere in the city. It was printed on beer glasses in bars, in clothes shop windows that announced "Forum sales," in museums, and on tour buses. It was a sui generis event tailored to make the most of the city's concerns for social justice, and it lasted for 141 days throughout the spring and summer. Three levels of government organized the event (the city of Barcelona, the Catalan Autonomous Region, and the Spanish state), and it also received strong sponsorship from the private sector. As an event, the Universal Forum of Cultures was radically different from any preceding sporting event or universal exhibition. Its organizing committee claimed that its purpose was to highlight the value of global cultural diversity and the need for more public awareness about social and environmental issues.

The organizers themselves had a hard time explaining what the Forum was about. They tried to sell the image of a fair of cultures. Its program included artistic events and representations from different parts of the world, such as indigenous dancers from Mexico, puppeteers from Vietnam, musicians from Brazilian favelas, Indian acrobats, and African drummers. Its program boasted a vast array of concerts, plays, shows, performances, and exhibitions. The Forum's image had been carefully crafted by several media and marketing companies, all of whom were partners in its overall running. The newspaper *El Periódico*, for example, had within its pages a countdown of days left before the event started as well as a whole section dedicated to news and photographs of the event itself while it lasted.

Despite the challenges the organizers faced defining their own event, they nevertheless stated its aims as threefold: peace, multiculturalism, and sustainability. The Forum clearly used the language of the global justice movement, blatantly copying some of its slogans. The fact that its organizers called it a "forum" led many of those involved in the FSMed to accuse it of taking advantage of the increasing popularity of the World Social Forum. Its blunt use of traditional contestation symbols—like an antiwar one—led the vast majority of social movements and alternative organizations of the city to oppose it. To blur the line even further with the World Social Forum, the fourth meeting of the Local Authorities Forum took place in Barcelona under the auspices of the Forum Barcelona 2004. Up until then, this gathering of representatives from city governments had been taking place as part of the World Social Forum.

Yet another way in which the Forum used the contested brand of Barcelona was with the idea of "dialogue." On May 9, 2004, *El Periódico* showed a reduced image of the paper's front page from February 15, 2003, the day of the massive, globally coordinated protests against the war in Iraq, titled "'No' a la Guerra" (No to War), with the antiwar symbol, beside a full-page marking the start of the Forum, titled "'Sí' al diálogo" (Yes to dialogue), with a photograph of the metal sphere that served as the centerpiece for the Forum's main performance, "Move the world." *El Periódico*'s editorial linked the strong antiwar sentiment in the city directly to the Forum: "With the same strength with which it came out on its streets to oppose an illegal and unjust war, Barcelona made a solemn vow yesterday to world public opinion in favor of dialogue. That was the message of the opening of the Forum. The speeches of the King, Juan Carlos, [the head of the Generalitat] Maragall, [the mayor of Barcelona] Clos and other institutional representatives provided a brilliant framing of the event, in which [the Spanish prime minister] Zapatero received a warm response from the audience, when he defended a concept of Spain that reflects all our plurality." With this editorial, the newspaper also linked the Forum's preoccupation with multiculturality to Spain's "plurality," which in Catalonia mostly means the recognition of its "distinctiveness."

Before the event had even started, there had already been a series of informative meetings, publications, and public condemnations. Leftist intellectuals used the Forum, organized by a private consortium with the government authorities of the city of Barcelona and the Catalan Autonomous Region, as a starting point for strong criticism of urban planning and policy making (Martí and Del Olmo 2004). An edition of the critical magazine *Archipiélago* was dedicated to the theme "Crisis and reinvention of the contemporary city." A book was put together in Catalan with articles from a wide array of local writers, titled *Barcelona marca registrada, un model per desarmar* (Barcelona registered trade mark, a model to dismantle), which was made available online (UTE 2004). The contributors' criticism centered on the fact that the governments of Barcelona and Catalonia (the latter of which had, at that time, recently come to be headed by the Socialist Party of Catalonia) sought to make the region a key economic player at the European level by promoting its port facilities and seaside location. Such a strategy would, according to Martí and Del Olmo (2004), help Barcelona avoid the deterioration that some industrial cities have suffered throughout the West because of the global trend of relocating industries to developing countries. These and other authors claimed that the Forum was only a vehicle to fund a regional transformation that would not benefit the majority of the population. Two other books strongly questioned the Forum: *La otra cara del Fòrum de les Cultures S.A.* (*The other face of the Forum of Cultures Inc.*) (Varios 2004) and *Barcelona 2004 como mentira!* (*Barcelona 2004 as a lie!*) (Trallero 2004). In an interview on the topic with a local newspaper, the sociologist Dominique Wolton said that "multiculturality is more than a festivity and a business" (Navarro 2004).

On the day this Forum held its opening ceremony, a protest took place in front of the purpose-built area where all the high-profile guests were arriving, which included the Spanish royal family, politicians, artists, businesspeople, intellectuals, and others. I was in the anti-Forum rally with Víctor and other members of the FSMed's organizing committee. We were all making noise, showing placards, distributing flyers with information about the corporations funding the Forum, and giving out pins with the slogan "Jo no sóc Fòrum" ('I am not part of the Forum'). Several of the flyers criticized the fact that the Forum's organizers claimed it stood for peace and social justice while some of its sponsors were weapons manufacturers. It was also criticized for its high registration fees. A single day entrance ticket cost €23 (US$28), which meant that the vast majority of the population could not afford it.

One of the Forum's main features was a large multimedia exhibition on languages called *Voces*, which was curated by the museum designer Ralph

Appelbaum. The exhibition included videos of people speaking hundreds of different languages with complementary information and maps showing where each language was spoken and how many people spoke it. It also had statistics on language use on the internet, on the media, and on music, as well as many other small displays that the curator thought could "raise awareness about the fragility of the linguistic diversity" (Fontova 2004). Another feature of the Forum was to bring together many already established annual events in Barcelona that receive government funding and to place them within the context of the Forum. Such was the case with the Grec Festival (classical Greek theater festival) and also with exhibitions in different museums. One of these, in the Fundación Antoni Tàpies, included pieces that were critical of the Forum itself, and so it was easy for the organizers to say that contestation was an essential part of the event.

The Forum's organizers also had the idea of endorsing 141 books related to the "core themes" of the Forum (the number was chosen to match the event's duration of 141 days). This was achieved through the placing of 125,000 stickers with the Forum's logo on the covers of the books of forty-six publishing houses in sixty bookstores (Andreu 2004a). One of the books was *No Logo*, written by Naomi Klein. The internationally acclaimed author and social activist was noticeably upset about this when she visited Barcelona on May 1, 2004, to launch her film *La toma*, about Argentinian cooperatives. "My book showed the new economic and symbolic value of brands . . . and the logo imposed on *No logo* is symptomatic of how the Forum has been set up," she complained to journalists (Andreu 2004b).

A group of NGOs was invited to be part of the Forum; they hosted fifteen exhibitions illustrating the United Nations' Millennium Development Goals (Blanchar 2004). The participating organizations included: Intermón Oxfam (Oxfam's Spanish partner, based in Barcelona), with a fair trade shop; Fundación por la Paz (Foundation for Peace), with an exhibition in favor of disarmament; and Médecins sans Frontières (Doctors without Borders), with testimonies from refugees. Those that took part in the Forum were but a small fraction of the number of organizations that exist in Barcelona. Numerous NGOs were invited but did not want to take part due to their leaders' skepticism about the event and its purposes. Those that did participate were seen by radical activists as collaborating with unjust practices that participants camouflaged as legitimate. Of the activists with whom I associated during my fieldwork, Javier, from the organizing committee of the FSMed, was one of the strongest critics of the Forum. He told me that there was a rumor among people in the Catalan government that officials had threatened some NGOs with reducing or withdrawing their official funding if they did not take part in the Forum.

There was an Assembly of Resistance to the Forum (Assemblea de Resistències al Fòrum), in which several of those involved in the FSMed took part. This assemblage was responsible for the protest during the inaugural ceremony already described and for several marches in the streets of Barcelona, including a *cacerolada* (demonstration where saucepans are banged) on May 8, 2004. Activists involved in this assembly wrote letters to high-profile activists who had been invited to the Forum to explain their criticism of it. They claimed to have convinced Günter Grass, Noam Chomsky, Naomi Klein, and José Bové to refuse to take part in it. "The Forum is an intelligently set up operation that tries to replicate through lies and a perversion of meanings the social movement forums that already take place," the economist Miren Etxezarreta told *El Periódico* (Tramullas 2004b). This assembly printed maps of the event with details on the sponsor corporations' unethical behavior.

The influential Federation of Associations of Residents of Barcelona published a fiercely critical article of the Forum in its magazine, *La Veu del Carrer*. It accused the government of using the Forum as a strategy to continue with the urban speculation trend that started with the 1992 Olympic games (Tramullas 2004a). It provided figures to substantiate its claims. An example is the comparison between the advertising expenses for the Forum of €50 million (US$61.5 million), compared to the rehabilitation plan of La Mina, a poor neighborhood close to the Forum, which was set at €78 million (US$91.1 million) over a ten-year period.

At the time of my research, the three levels of government in Barcelona, Catalonia, and the Spanish state were in the hands of the left, that is, the Socialist Party (in Catalonia it was a coalition of the Socialist Party of Catalonia, Esquerra Republicana, Iniciativa per Catalunya, and Els Verds). The government had changed only recently, as it was elected on December 16, 2003, and took office on December 20 of that year. Before the Socialist Party took over the Generalitat, the government had, for twenty-three years, been in the hands of the nationalist conservative coalition of Convergència i Unió, led by Jordi Pujol. One cultural event carried out in the context of the Forum that took center stage for the newly arrived government representatives was a theater version of Orwell's *Homage to Catalonia*, in Catalan. On May 24, 2004, the president of the Generalitat, Pasqual Maragall, and other ministers and local authorities attended the play's first performance (Barrena 2004).

Orwell's take on the situation in Catalonia during the Spanish Civil War has been considered an honest account of what happened in 1936 and 1937 in the region. He is earnest about what he saw as the Soviet meddling that ended the effusion and optimism he witnessed when first arriving in Catalonia in 1936 (Davison 2001). As he arrived, anarchists had abolished class stratifications, and Orwell's and other accounts speak about a generalized optimism and effervescence in the region. It was not long before the Soviet-backed International Brigades started Communist purges and reinstated privileges for some while instilling fear in the population. Orwell barely escaped a trial as the unit he had been part of was disbanded and accused of treachery. It was no surprise, therefore, that he developed an intense dislike of Stalinism, which drove him to write his eyewitness account of events as well as *Animal Farm* (Orwell 1945), a critique of the Soviet Union and Stalinism.

For a long time, Catalans remembered the positive atmosphere Orwell witnessed at his arrival in Barcelona. This reminiscence of solidarity and optimism inspired activists to challenge government policies that allowed for wealth concentration in a few hands. Throughout decades, such a principle also remained as an ideal for some Catalan ideologues. Once Franco took power, people were forbidden to speak Catalan in public and to show flags and other symbols of the region. Resistance to the central government was thus practiced in indirect ways. In a similar fashion as in Brazil, where Christian Base Communities (Comunidades Eclesiais de Base, CEBs) provided safe havens for dissenters during the dictatorship, in Catalonia, this role was undertaken by cultural or sport associations. Examples of this were the *Destino* magazine (Cabo 2001), the Football Club Barcelona (Cirici and Varela 1975), and the collective effort to build a museum in honor of Picasso for the commemoration of the hundredth anniversary of his birth (Vigil y Vázquez 1981). The soccer team's slogan, *Més que un club* (more than a club), alluded to its role as the custodian of the Catalan identity (Chadwick and Arthur 2008). All initiatives mentioned sought to establish links with foreign trends, enhancing a fruitful exchange between local artists or soccer players from abroad. These institutions helped provide continuity to the city after the fall of the dictatorship.

Especially relevant for this volume are two aspects of Catalan history that are often highlighted by nationalist movements: its early democratic character and its strong links to the rest of the Mediterranean. As regards its early democratic character, ever since the fall of Franco's dictatorship, various groups have worked to establish a nationalist agenda. Both aspects stand out as the most common references used by local political leaders for the legitimation of the region's claims to a place among independent states. During the Middle Ages, Catalonia was densely populated and was bustling with commercial activity (Farías Zurita 2009). As Barcelona grew in size and influence in the region, three types of institutions were developed that are considered by historians as some of the earlier signs of democratic decision-making assemblages in Europe. One was the Consell de Cent, which was established in the thirteenth century by the king of Catalonia, Jaume I (James I). Its duty was to elect three members of the municipal council that would govern the city of Barcelona (Florensa i Soler 1996). Another institution was the Cortes, the advisory council made up of the feudal lords. The third institution was the Consulate of the Sea, through which Barcelona merchants had the right to settle their commercial disputes without interference from royal courts "on the basis of a special law, developed out of practice" (Blockmans 2008, 16). It was also developed in the thirteenth century and was meant to oversee the city's commercial networks as they expanded around the Mediterranean.

This brings us to the second aspect: Barcelona's role in the Mediterranean. Once the Crusades, which began in the late eleventh century, had cleared the commercial routes in the Mediterranean from Islamic attacks, several maritime powers vied for their control. Bensch explains, "By 1300 . . . the Catalan capital had forced its way into the leading ranks of Southern European towns. Its merchants competed with the Genoese for economic domination of the Western Mediterranean, its municipal council supervised Catalan trading outposts stretching from Seville to Alexandria, and its financiers held lucrative administrative positions in the extensive dynastic confederation known as the Crown of Aragon" (Bensch 1995, 2). As part of the Kingdom of Aragon, it exerted control over Sicily, Naples, and Sardinia.

Such history was present in many ways at the Forum Barcelona. One of the main exhibitions of the Forum was on cities. It included bird's-eye views (or rather, satellite images) of dozens of cities. A few of them were shown in stages representing their growth over time. One of these was Barcelona, which started originally as a Roman military fort around the year 15 BCE (Castellar-Gassol 2000; Agustí 2002). The growth of the city followed the various waves of invasions and commercial connections the region was subject to during the subsequent centuries. The exhibition ended with a layout of the Catalan government's strategy to ensure Barcelona becomes one of Europe's leading ports, and the most important Mediterranean port. In this exhibition, as in the rest of the Forum, curators used history to promote regional growth in such a way that they appeared to ignore the claims to democracy and justice on which most of the rhetoric around the Forum's promotion was based. It did not contradict, however, the aim of using Barcelona's place in the Mediterranean to position Catalonia as an important territory in the region.

The tension between democratic ideals and pragmatic aims permeated numerous discussions I witnessed during the planning leading up to the FSMed event. Although the collectives organizing the FSMed claimed their purpose was to facilitate a regional meeting that would provide opportunities for civil society groups to network and address some of the region's problems, they often benefited directly from meeting and negotiating directly with government officials. By claiming to organize a Mediterranean-wide event,

Barcelona-based groups gained legitimacy that, in the eyes of federal, state, and local authorities, made them bridges to the region's civil society. Many of the struggles that took place among the radicals and the social democrats could thus be explained as both camps vying for gaining that legitimacy. Although some in the radical group were quite critical of Catalan nationalism, in practice their own work for the FSMed supported its claim to offer hope for the region's civil society.

In this chapter, I have offered an overview of the activist and advocacy milieu in Barcelona at the time of fieldwork, especially in light of the dominant nationalist agendas that pervade major events. Using an ethnographically informed revision of the history of progressive movements and events, I have provided valuable information to understand some of the pervading tensions among the groups that participated in the organization of the FSMed event. To the differences between radicals and reformers, which are similar to those in Brazil, here we can notice that some members of both share their progressive ideals and practices with nationalist agendas, although most seem to be reformers. Since the time of fieldwork, less attention has been placed on the city's progressive credentials as the dominating debates in the public sphere have revolved around nationalism. In the coming chapters, my analyses of advocacy networks as cases of civil becomings explore the constant tensions and frictions activists and advocates need to deal with.

PART TWO

Plural Networks

3

Advocacy Networks' Communicative Characters

Networks allow you to have a wider vision, that is, above all else
for organizations like ours, which because of the very specific
scope of union work, we run the risk to end up with a corporatist and closed vision . . . so networks give you a global vision
that allows you to incorporate to your daily endeavors a series of
experiences and vindications and allow you to connect to large
problems and large policies what is happening in your workplace or sector, which for us was very necessary.

— Javier, union leader and member of the FSMed Technical
Secretariat

SOON AFTER I arrived in Barcelona, Víctor invited me to a meeting of the
Mediterranean Social Forum (Fòrum Social Mediterrani, FSMed). In his role
as coordinator of Socis de la Terra (SdT), he was keen for me to find an issue that could be the focus of my volunteer work. "I can't be everywhere, so
I need someone to sit there [in the FSMed organizing committee] and serve
as a representative of the SdT. If you do it, you can tell me what happens and
we can agree on how to go about things," he told me. We had been in touch
for months prior to this while I negotiated access to the organization for my
research, which SdT was pleased to offer me in exchange for my work as a
volunteer. The SdT was involved in several projects, mostly liaising with the
local government of Barcelona or the regional Catalan administration, for example, those promoting organic food in school canteens or advocating for a
reduction in the use of pesticides. Because of my interest in the World Social
Forum process, I decided to stay with the FSMed.

This chapter is an exploration of the communicative characters of advocacy networks. If any social assemblage that refers to itself as an organization
relies on communication to maintain the social ties that bind it together, then

a network of such organizations needs this to a larger extent. What is argued here is not the stated purpose of advocacy networks in generating new public spheres, where all members of political communities can engage in common issues (Castells 2008), but rather the quotidian practice of communication through which ideals, ideas, plans, practicalities, and problems are defined, shared, and negotiated in order to reach common decisions for collective action. Communication is thus the infrastructure that is required for action, which will be analyzed in the form of entangled agency in the following chapter. It is also at the basis of the performances I argue lie at the heart of the project of civil becomings, that is, of how groups of individuals enact their membership of civil society organizations as an exercise of legitimacy to address public issues. For Keck and Sikkink, networks can be understood as "communicative structures [or also as] political spaces, in which differently situated actors negotiate—formally or informally—the social, cultural, and political meanings of their joint enterprise" (Keck and Sikkink 1998c, 3). This means that networks may be fragile: they can easily fizzle out if participants do not feel that they are shaping the emergent assemblage; or, they may become resilient, if the opposite happens.

I therefore stress the communicative as what enables advocacy networks' plurality at work. One of the leading members of the FSMed organizing committee, Javier, explained to me once in his office that being involved in transnational collaborative efforts had widened the horizon of the independent union he led: "Networks allow you to have a wider vision, that is, above all else for organizations like ours, which because of the very specific scope of union work, we run the risk to end up with a corporatist and closed vision . . . so networks give you a global vision that allows you to incorporate to your daily endeavors [*acción diaria*] a series of experiences and vindications and allow you to connect to large problems and large policies what is happening in your workplace or sector, which for us was very necessary." This, of course, is achieved through communication. In particular, this chapter will focus on the fieldwork in Barcelona and its context.

As I became immersed in the FSMed organizing process, I realized it had been plagued with difficulties. Other chapters of the World Social Forum process were focused on either cities, countries, smaller regions, or unifying topics. The FSMed was the first regional forum to seek a bridge between affluent influential countries (Europe) and conflictive regions (the Maghreb, the Mashriq, the Balkans). On several occasions, I asked some of the activists about the FSMed's history. Javier, a trade union leader who was a good friend of Víctor's, told me that Egyptian trade unions had kick-started the process but had dropped out after they felt that it had been overtaken by European groups: "Other Egyptian groups have taken over in that country, but the original proponents have distanced themselves from the FSMed." Conflicts were so prevalent that the date for the meeting had been postponed twice due to difficulties among participating groups. The challenges of a regional forum like this were monumental. When I asked Javier why the FSMed process had been so difficult, he inhaled deeply from his cigarette, exhaled calmly and then started explaining:

> There are basically two reasons for the ongoing delays: first, is that we who initiated this process maybe did not sufficiently estimate the problems in relations, in communication, that the different cultural sensitivities would generate. We were used to dealing primarily with European and some Latin American groups, but mostly European [. . .] and we thought that we would find movements that were consolidated and ready to negotiate as we were used to. But in reality, it is not like that. The political culture in the south [of the Mediterranean] is very different from ours. It needs more time. Processes mature more slowly. There is a certain distrust against the north—absolutely justified, I think—and so it has not been easy to find organizations and movements in the south that would get involved in an intense and open way in the process. More time for mutual reconnaissance and discussions has been and still is necessary, in order to allow for people in the south to see the project as something they own, like something where they are protagonists as we, the movements from the north, can be. [. . .] And the second reason I think is because of some people in the north—not those of us in the collectives and organizations who have been involved in the process, but those others who have not been part of it—they have had two motives for not becoming involved: because they did not trust that this process could consolidate and advance, and also I think that many colleagues [*compañeras y compañeros*] of the social movements who are actively participating in the preparations and in the European Social Forum, I think they saw it as something opposed to the European forum that could damage it and attract some of its same participants; it could distract efforts and thus break or weaken the European Social Forum.

The frictions that occurred in the networks were due to a series of presumptions regarding interactions and the type and style of activist groups outside Europe. By assuming that communication would flow with activists from such contrasting sociocultural contexts as it did among Europeans, FSMed organizers appeared to have little experience in interactions across cultural divides. The fact that Javier referred to "cultural sensitivities" means at least that there was some sort of understanding, although in my view very limited. Despite knowing of their shortcomings and challenges, for example, I never knew of FSMed organizers requesting assistance from anthropologists or other specialists with more rounded knowledge about such "cultural sensitivities." It appeared to me that FSMed organizers relied on their own contacts and personal experience with activists from southern countries to make do. I noticed how Víctor, for example, was more aware of possible misunderstandings, perhaps due to his Argentinean roots and his experience as an immigrant to Catalonia. The starting dialogue in a meeting with activists from around the Mediterranean summarized the frictions and problems that happened frequently. María seemed upset when she posed the first question: "I ask our colleagues from the south to explain what work they have done since our last meeting in Cyprus." To this, Víctor immediately replied in an annoyed tone: "And those from the north should do so as well." Everyone in the room nodded in agreement. One Algerian participant clarified: "It's not about those from the north telling those from the south what to do, but for both groups to talk about their problems. If we don't see it as a dialogue, then I am not interested in taking part in it. The FSMed has to look for its own identity, for its purpose. I need to know how to explain it to the grass roots to grab their interest. A woman who must walk miles to get a bucket of water is not interested in talking about imperialism." Despite frictions like these, organizers achieved some progress and managed to bring the FSMed event to fruition.

In also taking for granted that activists in dictatorships and less affluent contexts would be as "consolidated" as organizations in Europe, as Javier said, FSMed organizers proved to be extremely naive if not outright ignorant of other contexts. The fact that Javier could explain to me the effects of both sets of preconceptions so clearly, however, showed a level of awareness of communicative processes in the networks. In both cases, the communication failures were not only about content, or what was said, but were also due to contrasting expectations of bodily practices. Face-to-face presence in meetings, assemblies, and marches was crucial to earn trust and understanding between participants.

My interest in examining the communicative characters of advocacy networks is derived from the constant coming and going of information, opinions, deliberations, and decisions. In the many gatherings, meetings, marches, and events with activists, I was able to identify what I now classify as five communicative characters of advocacy networks, which I name as: diagnosis, conveyance, deliberation, resolution, and proclamation. This classification is not a clear-cut separation of activities or purposes but rather has the analytic goal to distinguish dispositions and practices that are building blocks of the relations formed through networks. In some cases, meetings and other gatherings may focus on one of them, but it may also be the case that several take place in one single meeting. I explain each of these characters in more detail further on, especially in relation to specific encounters where something is exchanged, be it information, opinions, analyses, ideas, or plans. These interactions are by no means purely reasoned processes, as Jürgen Habermas (1981) set out in his communicative action theory, but are rather punctuated by "lifeworlds" made up of personal emotions, cultural sensitivities, historical legacies, and affective reverberations of various sorts. This approach is thus markedly different from social movement scholars' manner of defining communication as either a sociological exercise of framing or as part of a political process of agenda setting. At the time of fieldwork, social media did not play a role in links between activists or in their campaigns, as is the case nowadays. Nevertheless, I believe the following analyses may be relevant to understand similar processes that take place among—or because of—activists.

In Barcelona, I spent my time between SdT, the FSMed, other campaigns and events that activists invited me to, and a few external events I considered relevant (like the Forum Barcelona I referred to in chapter 2). Some of these were cyclical processes to update information but also to recapitulate the relations between those attending. At SdT, for example, we would hold weekly meetings every Monday, where Víctor would ask each of us volunteers to report on what had happened the week before and what was to come in each of our areas. At the FSMed process, we would meet regularly (at some points it was weekly, at others more often), in different meetings and events. I also attended sporadic marches, talks, or other events that activists would invite me to. Each of these scales of participation were used by some to confirm one's role in and/or commitment to the wider movement, in a specific committee, or a cause. This happened not only with one's presence, although that was part of it, but also in one's participation, dialogues, debates. It was often the case that after any of these gatherings a group (or several) would carry on talking over a beer in a bar close by. This extension of activism as part of social life included comments going into personal details that would not have been shared in the more formal meetings or events. That's where I found out about a torrid romance between two activists who in meetings appeared to be on opposite sides of an ideological spectrum, or about long-held distrust among radicals of some reformist NGO workers. Such forms of socializing and gossip helped establish a sense of familiarity by asking more personal questions about political viewpoints or other aspects. Because they relied on seeking to establish a basis for collaboration, I include them in the first communicative character, of "diagnosis."

Diagnosis: Of Talk Shops and Forums

"We talk a lot, not so much to convince but to make people think," Víctor said in one of SdT's Monday meetings. He was referring to an event at a local university where he had spoken about pesticides. Like many NGOs, SdT worked mostly with communicative strategies, managing information and analyses with which its members strove to convince government officials to adapt policies in order to avoid identified problems or risks. I noticed that at the root of these campaigns and strategies, however, was the claim from NGOs or activist groups to have identified a problem or issue that had not been identified before or to draw attention to a problem or issue whose importance had been relegated by others. This is what I mean by "diagnosis." In my view, much of the work of networked activism is related to a mutually legitimizing form of diagnosis. SdT, for example, joined forces with other NGOs and international organizations to lobby the local government in Barcelona to offer organic and locally produced food in school canteens. By coming together and agreeing on the same issues, each organization brings in its different viewpoint, which may include complementary information, considerations, or histories. Agreeing on a diagnosis is crucial for any network that considers collaborating, as it allows for mutual understanding in interaction. It has been perhaps part of the learning process of activist groups to allow for spaces where there is enough openness and lack of structure for dialogues to result in agreed positions. This is what I consider to be the case with the World Social Forum (WSF), which is the umbrella organization under which the FSMed took place.

One of the most common criticisms of the WSF over the years is that it is a "talk shop," or a place where activists simply get together to talk, often as much venting their frustrations as stating their dreams of change (Whitaker 2002). In January 2003, I attended a WSF event for the first time. It was in Porto Alegre, Brazil. As I walked into different tents, I was reminded of academic conferences with a myriad of parallel panels and a few massive gatherings for keynote speakers. Forum organizers stress that activities are arranged by participants. The WSF is considered a process because it promotes dialogues among different activist groups for them to put together activities (which can be panels, workshops, exhibitions, performances, or other) that will be held during the WSF event over a few days. Panels and other activities were thus convened by activist groups, often to discuss one or several problems and to share their projects and plans that were considered as part of the solution. The key difference between social forum panels and academic ones was in the participation of the audience. After those in the panel had spoken, a microphone would make the rounds for questions and comments. But these exchanges were less about asking for clarifications than about sharing what others were doing. Whenever someone would take the microphone, they would start telling the story of their own activism, their experience trying to convince government officials or maybe their own neighbors about the need to change policies or everyday practices. These interventions often lasted a few minutes. They seemed to constitute rituals of inclusion, where participants felt heard after perhaps having faced difficulties by themselves without many supporters. In January 2004, I attended the WSF in Mumbai, India, which, although more vibrant in the hallways than in the meeting rooms, had a similar air of excitement than the one in Porto Alegre. At one point, while having dinner from dried-leaf plates in one of the rest areas, I met Sandeep, an Indian activist who told me that he was very happy to be there: "I have worked for a long time in a small community, almost alone, and this," he said, while extending his arm and pointing around us, "meeting so many people who want to know about my work, and who congratulate me and give me energy to go on . . . is very beautiful. . . . I did not expect it." Sandeep shared with me his sense of having found a wider community of like-minded people, one not based on sharing an identity but rather on a mutual solidarity and support.

For many intellectuals, the World Social Forum has helped provide a new backbone for a global left after the implosion of the Soviet Union and its peripheries (Santos 2006). It emerged out of the loose activist network called People's Global Action (PGA) and held its first meeting in Porto Alegre, Brazil, in 2001. It was designed to symbolize an outright opposition to the World Economic Forum, which brings together some of the most powerful corporations and individuals alongside states and international organizations (Fisher and Ponniah 2003). What defines the WSF is its network ethos and a sense of optimism captured in its slogan "Another world is possible" (Acosta 2009). Its promoters insist it is an arena for dialogue above anything else. For some of its high-profile participants, the momentum gained in the meetings required its transformation into a political organization in its own right, similar to a global political party (Monereo, Riera, and Valenzuela 2002). They complained that the whole event ended up being just a place to "talk shop," and that it failed to take advantage of the political capital it could have gained in bringing so many different organizations together. In contrast, the prevailing position has been to maintain the ethos of the WSF as a meeting place or an umbrella organization, without seeking to build up a political force in its own right (Whitaker 2005; Santos 2005). "Everything we do," a Canadian NGO manager I met in Porto Alegre told me, "is in collaboration . . . we need others to do even the tiniest thing." He was referring to the work of his organization in partnership with local groups around the world. The WSF itself is known as a "network of networks" (Santos 2005), which in turn promotes the emergence of new networks (Fisher and Ponniah 2003). It was precisely the open and unencumbered communicative character that allowed for perhaps unforeseen collaborations to take place. This creative force through open-ended communication drove my interest in the process. I noticed time and again that in many of its sessions, some activists took the microphone and spoke for longer than would have been necessary as if they had a need to be heard. In talking openly about experiences and problems, there seemed to be a search for a way of naming a predicament that would more easily unlock a potential solution. But they were also in search of collaborators, of teams, and of inspiration. Diagnosis therefore constitutes a search through dialogues as a basis for collaboration.

Conveyance: Clarifying Concerns, Demands, and Proposals

While activists may agree on considering something a shared problem or grievance, they usually need to trust one another in order to collaborate. Within established groups, trust has been built by membership, acquaintance, and fraternization, which make up acts of performative allegiance (Horton 2003). In networks, such trust comes partly from what I call "conveyance," or to clarify one's concerns, demands, and proposals. Such clarification is also a way to expand the sense of community with whom activists wish to fraternize. In building associations with others, activists expand their sense of belonging. This is informed by Eriksen's view of anthropology, which "distinguishes itself from other lines of enquiry by insisting that social reality is first and foremost created through relationships between persons and the groups they belong to" (Eriksen 2004, 9).

As part of the preparations for the FSMed, several participating groups organized a seminar at the Centre of Contemporary Culture of Barcelona (CCCB) in March 2005. The event, titled "Ten years of the Barcelona process" (Europartenariat), included three renowned speakers analyzing the economic, political, and social history and implications of the European Union's deals with regional governments, which started in 1995. "After the fall of the [Berlin] wall, the Mediterranean was forgotten. The European Union's priority was the East," explained a local law professor. He then described how it was business leaders who sought favorable legislation in southern countries for increased trade and investment. Issues regarding migration, remittances, debt, privatizations, human rights, and terrorism, among others, were discussed, mostly as they pertained to North Africa and the European Union. The panel was very well attended and was a helpful way to publicize the upcoming FSMed event. In their talks and the ensuing discussions, I heard constant reminders of the region's long past as well as plenty of ideas about potential futures. Academic analyses provided ideas and elements to activist groups and NGOs for their campaigns and collaborations. There was a clear consensus in that room, as in many of the spaces of the Forum, to promote a progressive agenda.

In some gatherings, each group's position on an issue was met with skepticism, critiques, or outright dismissals from others. But the fact that they were there, listening, was a first attempt at collaborating. "Let's try to understand each other!" claimed Javier once when trying to calm nerves down after a row erupted in a meeting. In the study of activist networks, there is sometimes the tendency to look for coincidences or perhaps parts of the networks where consensus already exists. It makes sense to seek as coherent a community as possible to define it and study it accordingly. Sometimes, the path to finding such a sense of community is research as an extension of one's own activism, as is the case with Juris, Graeber, and Maeckelbergh, who have carried out studies on complex activist networks using their own activism as an entry point (Juris 2008a; Graeber 2009; Maeckelbergh 2009). Juris focuses on the "anti-corporate globalization" (Juris 2008a, 9), Graeber on the "radical direct action" (Graeber 2009, x), and Maeckelbergh on the "alterglobalisation movement" (Maeckelbergh 2009, 33). Their studies are thorough explorations of radical activist collaborations, from which mine markedly differs on two levels: I am no activist, and I do not seek to highlight a coherence in their collaboration. While I sympathized with many of the critiques expressed by numerous WSF and FSMed participants, I avoided their dichotomies of good (activism) against bad (states, capitalism), as they appeared to be fixated on a static understanding of purity and danger. What caught my attention even in my earliest exposure to the WSF was the possibility of dialogue between all the different left-wing groupings, especially considering they aspired to influence a wider social assemblage than that with which they shared ideologies or identities. From my perspective, therefore, the plurality of views and of habitus among all those who were gathered at the World Social Forum demonstrated in itself a willingness among the participants to deal with challenging differences. In the FSMed organizing process, it was clear that what was being built was far more diverse than a single ideological viewpoint. But this was precisely part of the challenge.

Deliberation: Dialogic Negotiations

To seek common ground, activists must go beyond simply sharing their points of view and attempt to consider those of others. This was a constant struggle in the FSMed organizing process, as there seemed to be cases of people simply talking over each other. "To achieve any meaningful change, we need to convince people in power, to persuade them," Víctor told me over a beer after the seminar at the CCCB I mentioned previously. The persuading, I noticed, was also necessary within the networks, where different activist groups try to convince one another that their way of doing things is ideal, or their viewpoints are best. Only after there is some type of consensus can networks negotiate with local and regional government offices and other institutions. In a meeting held on April 11, 2005, just over two months before the FSMed event, organizers were discussing the solidarity fund, that is, the amount of money dedicated to help activists from southern countries to travel to Barcelona to attend the FSMed event. After months of negotiations with government authorities, it was not yet clear how much money FSMed organizers could actually spend to help activists from poorer countries attend. The debate thus turned to how to communicate this fact to all participants:

> José: Our budget is the Achilles' heel of the project (FSMed). We have created dependencies by paying the flights [of south-based activists]. We cannot show all our budget, because it will create expectations that we are swimming in gold. For this reason, we should not include the infrastructure contributions that the [government] institutions will make.

> Sergi: I vote that we be transparent and say the truth: that we do not have a single penny (*no tenemos un duro*). And we don't know if the [Mediterranean Social] Forum will take place, because we do not have certainty about what the [government] institutions will contribute. What happens if in May the [government] institutions only donate a quarter of what they committed?

> Javier: I think that we need not be optimistic, but also not alarmist. We have already been transparent in what we have agreed: that [government] institutions have committed to paying for infrastructures. We do not have any clarity of the solidarity fund. I agree with José in making public what we will spend as Forum, not the total, because it may lead to confusion. We have the spaces guaranteed, but not the solidarity fund.

The dependencies that José was referring to were due to the solidarity fund during the organizational process, for attendance to International Assemblies and the International Coordinating Group. For these trips, the funds used were from the Fons Català de Cooperació al Desenvolupament (Catalan Fund for Development Cooperation), where the committee had been holding its regular gatherings in 2004. The infrastructures that José and Javier mentioned were facilities of the Fira Barcelona, an exhibition area for trade fairs that is jointly run by government and private investors. In the spreadsheets used to convince government officials to increase their contributions, FSMed organizers would include the infrastructures with the price the Fira would usually charge. These were the numbers that were to be hidden from participants. The short excerpt I included was of a dialogue that went on for a considerable time. It reflected the careful balance that had to be managed by organizers between contrasting expectations: their own, government officials', and participants'. It also revealed the fragile state of finances for the FSMed event, which relied heavily on government contributions that in turn were estimated according to the number of attendees the organizers were able to convene. But what I found intriguing was the constant effort to control how others would perceive the FSMed: not as too affluent, for participants; and popular among activist groups and NGOs, to government officials. The way to get higher contributions from government institutions was to commit to bring a certain number of participants. A number that was constantly mentioned in 2004 was 15,000, which went down to 5,074, as organizers included in the final report (although this was actually higher than the numbers I had seen). Interestingly, the dialogues that took place in so many meetings preparing for the FSMed event were to facilitate more dialogues among a wider number of activist groups for the event. I believe that the difficulties that organizers had resulted in a much reduced number of potential beneficiaries from the Forum as a space for dialogue.

Throughout the organizing process, however, the solidarity fund had been used to pay for activists' travels to international meetings and assemblies. Conflicts often arose over who would benefit from such grants. Many organizers clearly sought to benefit their close partners in the south (called "sister organizations"), even if some of these partners already had the possibility of using funding from other sources. One of the meetings where this was discussed in 2004 lasted over five hours and was full of shouting and mutual accusations. A week after the meeting, while having a beer after another meeting, Rosa told me that "their strategy is to tire others out, so that everyone needs to leave and then they make the decisions." It was true that after a few hours, several people left, and eventually only very few remained, among whom it was easier to negotiate. Rosa was nevertheless indirectly accusing María and her reformist camp of trying to favor their partners, while implying that Javier, from the radical camp (to which she belonged), had stood his ground in order to at least even out the invitations for other groups. Dialogues were therefore often full of tensions, as it was not pure will to listen to others, but rather a sense of political negotiation that took place. I differentiate these exchanges from actual decision-making, which has a character in itself; I call it Resolution.

Resolution: Who Makes Decisions in a Network?

To decide on common action or positions, a network usually has already agreed on the form that such decision-taking should have. Perhaps because of the bureaucratic experience of some network members, a few rules or

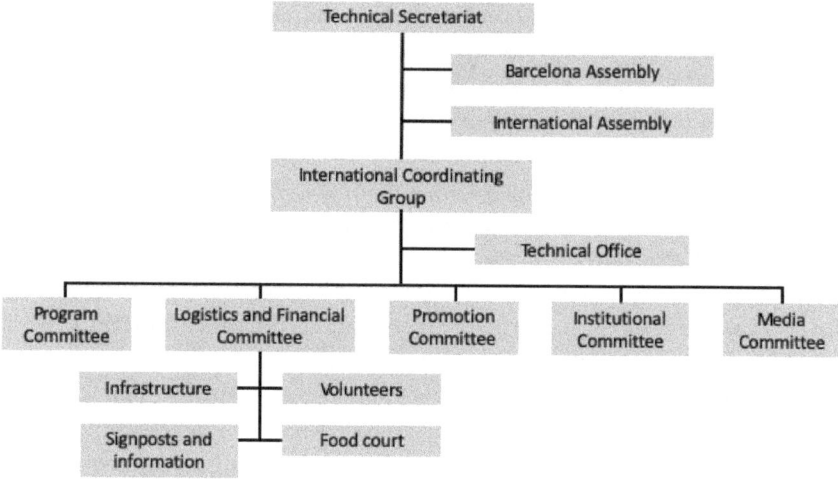

Figure 3.1. Official structure of the Mediterranean Social Forum (FSMed) organization committees in 2004. (Raúl Acosta)

procedures are set forth. In the case of the FSMed, there was a whole structure that underpinned different levels of decision-making in the organizing process, or so I was made to believe. In many meetings, I kept hearing about assemblies and meetings without understanding much. In one of those after-meeting drink evenings at a local bar in central Barcelona, I asked Javier about this. He started explaining more clearly the various responsibilities of the different assemblies and closed groups involved. The organizational arrangement, he told me, was inspired by that of the WSF, which placed particular weight on assemblies as guiding decision-making bodies. Because no chart existed about this, I sketched one according to what Javier and others told me (figure 3.1). I use it to analyze decision-making because it illustrates that there is often a tension between claimed structures of legitimate decision-making, and actual ones. It didn't take long for me to realize that this version of the organizational chart was not how decisions were made for the FSMed organizational process.

Charts of this sort attempt to conceptualize the distribution of power within an organization, the shape of which determines who gets to decide what. FSMed promoters claimed that the highest authority of the network was an "international assembly" that needed to vote on all important decisions. This assembly was open to all progressive groups in the region. It represented the "body politic" of the Mediterranean Social Forum, providing a degree of democratic legitimacy by allowing all interested groups to vote on its process. There were only a few meetings of this international assembly, each one in a different city (Málaga, Rabat, Istanbul, Marseille, Barcelona), over a whole weekend. It was not easy for many groups to send representatives to all meetings, but some could manage to do so. In its official documents, such as its final report after the meeting (FSMed 2006), the Technical Secretariat clarified that the key decisions were made during the International Assemblies. But, from what I noticed, the difficult negotiations would take place beforehand; during the assemblies, the results would be presented as only needing the approval of those present.

As interest in organizations has expanded in anthropology (Hirsch and Gellner 2001; Wright 1994), so have the calls to pay attention to their output in the form of policies (Shore and Wright 1997) and documents (Hull 2012; Riles 2006) as well as in their inner workings, for example, in meetings (Brown, Reed, and Yarrow 2017; Schwartzman 1987). As with other social forums, the FSMed organizing structures constituted a voluntary arrangement through which participating groups could justify decisions and actions. The chart I drafted above reflected the stated intentions of participating groups as they navigated negotiations among what they considered their constituencies, that is, the larger pools of activists. Like the WSF, the FSMed fell into a methodological nationalism by considering the participating groups from each country as representative of the country's progressive activists. Organizers knew very well that the Mediterranean was a difficult area because of the long-lasting conflicts and frictions that occur on its shores. Even the meetings held in different countries around the Mediterranean so that groups from the area could better participate were often contentious. Visa problems for activists were mainly due to the fact that those who had visited Israel, for example, could not go to certain Arab countries, and that some southern activists were refused visas for the Schengen area of European countries. This meant that only those groups with either enough funds or support from the organizers were able to attend. Perhaps the most difficult challenge, however, was to make decisions in a legitimate manner, without ostracizing or alienating other groups.

In their references to the ideal organizational structure, FSMed organizers claimed that the International Assemblies had the ultimate authority over the organizing process. In meetings of the Technical Secretariat, organizers often reemphasized this hierarchy with phrases such as: "As was decided on the Assembly in Cyprus" (María) or "After a long debate, the International Assembly made the decision in Casablanca of . . ." (Javier). Four of these International Assemblies were held before the FSMed event, and one more during it. The second most authoritative body for the FSMed process was the International Coordinating Group, which comprised a smaller group of delegates from each country. In the third place was another assembly, this time of the host city: Barcelona. Next in line was the Technical Secretariat, composed of a small group of local organizations in Barcelona who dealt with the practicalities of the FSMed. Specialized committees commissioned by the Technical Secretariat carried out specific tasks in preparation for and during the event, such as logistics or designing and printing the program. As administrative support, the Technical Secretariat set up what it called a Technical Office with two full-time employees and other volunteers to deal with more delicate paperwork (for example, visa applications from foreign activists).

This whole setup, however, was more ideal than real. Due to a lack of funding, I was unable to attend International Assemblies that took place outside of Barcelona. From what several activists told me, however, these were not very efficient as they tended to be dominated by local activists from each city where they took place who had not attended previous assemblies. This meant that issues that needed to be resolved quickly got stuck in cycles of clarification and repeated objections. The local Barcelona assembly I attended was quite similar. Many of those activists attending needed so much information about any decision that had already been made that it was not really possible for the assembly to make decisions that were required quickly. The International Coordinating Group was made up mostly of individuals with experience in the WSF process, but they did not meet very often. In reality, therefore, the actual decision-making structure was dominated by the Technical Secretariat. Both assemblies were used more as opportunities to share information about the process with like-minded organizations. The International Coordinating Group was influential, but more as a general outlook steering committee that could simply not handle the complications that arose during the ongoing organization of the FSMed. For these reasons, I developed a second organizational chart showing the power structure as it appeared to me (figure 3.2).

In the case of the FSMed process, therefore, decision-making was centralized in the Technical Secretariat, which was made up of activist groups and NGOs from Barcelona. This state of affairs was the source of some of the tensions and conflicts between the participating networks that plagued the FSMed. The difficulties that groups had with each other simply put an end to the possibility of a second edition of the FSMed event.[1] This is a good example of a working network, which achieved its stated goal, in this case of holding the FSMed event, but which because of its operation failed to keep the torch going for further collaboration. Decision-making is thus not only about the decision at hand but also about the legitimacy of the power structure

1. Ten years later, in 2015, there was a global edition of the World Social Forum in Tunisia, as the network sought to build on activist efforts in the Mediterranean after the so-called Arab Spring.

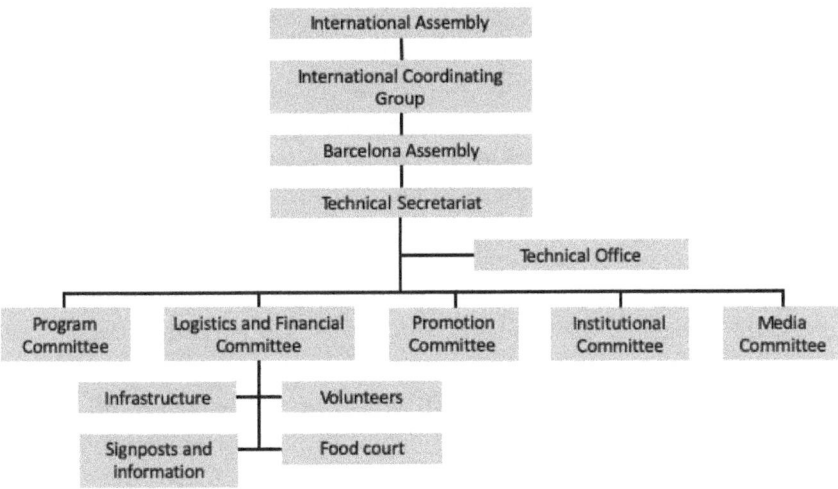

Figure 3.2. Actual decision-making structure of the Mediterranean Social Forum (FSMed) in 2004. (Raúl Acosta)

or agreed form to reach resolutions. I also differentiate this communicative character from that of public pronouncements of what a network seeks or does, which I define as proclamation.

Proclamation: Public Declarations of Common Aims and/or Activities

My experience with the Forum's visual identity proved to be quite telling not only of the difficulty of reaching agreements among FSMed organizers, but also of the basic understanding that all shared regarding their joint aims. Although not the only or best way, a network's chosen graphic design can help analyze its public statements of unified positions regarding its aims or activities. In the case of the FSMed, I found myself in the role of graphic designer to help bring about its visual identity. At the time of my arrival, in 2004, a professional designer had already created the FSMed's logo. It was a basic calligraphic design, which made it distinctive. But there was more to be done: "We need to do flyers, T-shirts, pins, and other material to promote the Forum [FSMed], as well as to ensure that the signposting within the venue is clear to those attending," explained Javier after I volunteered to do some further design work. Víctor had already committed SdT to being in charge of the design and the printing of signaling and informative posters within the FSMed, so adding promotional material seemed logical to everyone. "I am no designer," I warned them, but because of my experience with the relevant software they all agreed to let me do the work. As part of the FSMed visual identity, organizers wanted to include a slogan that would somewhat express its aims and scope. This debate was not as difficult as others, although it went on over several meetings. When a proposal finally met the approval of all, it referred to rights: "Mediterranean: a sea of rights." It was the more political alternative, proposed by the radical wing of the FSMed organizers. When Javier promoted it in the final vote, he clarified: "To say 'a sea of rights' emphasizes the aim, the hope, and the struggle . . . we know that in many places there are no rights, but everywhere there are expectations for rights, and the more we make them visible, the more we emphasize the positive, as does WSF's slogan of 'another world is possible.'" Once this had been agreed, a discussion followed about how many and which languages to include in the visual identity. Some were de rigueur, such as Catalan, Spanish, French, English, and Arabic. Italian was added, and so were Romanized versions of Arabic and Kabyle (a Berber language). The challenge I faced was to integrate all the languages into a single image. I thought of making each phrase a wave, and designing an abstract boat as the FSMed's symbol (figure 3.3). People liked it, and I had the green light to go ahead with printing all the publicity materials. Regarding the languages, in May 2005, once everything was printed, and just weeks before the FSMed, the representative of the Greek movements complained that Greek was not included among the languages in the visual identity. "How can we be left out? Our cultural heritage is key to the Mediterranean and our movement is strong," he complained in a meeting of the international organizing committee.

The visual identity that ended up being reproduced on T-shirts, stickers, posters, advertisements for newspapers and magazines, and other products thus captures the flux of interactions around what is also known as the "middle sea." For advertisements and flyers announcing the FSMed, a longer slogan was agreed on: "Struggle for a sea of peace and rights" (in Catalan: *En lluita per un mar de pau i de drets*). "This is to let everyone know that we are anti-establishment," Javier said in defense of the slogan during one of the meetings. Led by him, the radicals succeeded in pushing for the feisty slogan despite the qualms it raised among NGO people.

As part of my work in advance of the Forum, I also designed three large banners, each of which was twenty meters wide and one meter high. Each one had a different slogan in Catalan and Arabic highlighting three of the main thematic areas. The three selected slogans were: "Struggle for a sea of peace and of rights," "No to war, no to occupations," and "No to neoliberalism, no to patriarchy." The plan was to hang them in the auditoriums where the plenaries would take place, and at the end of the Forum, we would take them down to carry during the closing march. "This way we have two uses for them," Javier explained to me when I started designing them. As the Forum

Figure 3.3. Mediterranean Social Forum (FSMed) banner design with the logo and the waves above the entrance of the Fira de Catalunya at an FSMed event, 2005. (Raúl Acosta)

was drawing to a close, and it was almost time to join the march, a conflict erupted in one of the common areas between Moroccan and Saharawi activists. Numerous Catalan NGOs and activist groups have deals with Saharawis, as they are considered a national group that is threatened by the Moroccan state. The fight was thus part of a long conflict with which everyone involved in the FSMed was familiar: many Moroccans consider the Saharawi territories as part of Morocco, and not independent. As activists started shouting and fighting, Javier and others hurried to separate them. At one point, he asked me to fetch the banners since the march was going to start earlier than planned. In the end, because of the conflict and the ensuing chaos, we only had time to take two of the banners out to the streets, one to be held by those at the front, and another for somewhere in the middle. The march itself was uneventful, as local activists made sure to keep the contentious Moroccan and Saharawi groups separated. As has become common in such marches, there were drums, props, colors, slogans, and lots of enthusiasm. After a couple of hours of marching around a preestablished route that had been approved by city authorities, we all came back to the Fira Barcelona, and there was a final concert to celebrate the end of the FSMed.

The visual identity and slogans are examples of how advocacy networks make public statements about their purposes or activities. Other examples include press releases, messages in mainstream and social media, public statements, and similar. As was the case with the visual identity and the slogans, for such pronouncements to have legitimacy for network members, they should be the result of negotiations or conversations among its members, among whom there should be no strong objection. Because of their public character, these forms of stating a network's mission or program also help shape its sense of identity. They end up being messages as much for external actors as for network members.

This chapter explored the communicative character of advocacy networks, using some of my analyses from the Mediterranean Social Forum's organizing process. As assemblages of groups working toward a concerted goal or principle, advocacy networks take part in what Tarrow (1998, 18) terms the "political process." They do so, however, as much toward external authorities (e.g., governments or international organizations) as toward inside the networks themselves (among the participant organizations and groups), and also toward other stakeholders (e.g., grassroots organizations, urban dwellers, farmers, small-town inhabitants). This takes place through a complex interweaving of communication strategies and processes. I classify these as five communicative characters of advocacy networks: diagnosis, conveyance, deliberation, resolution, and proclamation. The interactions that occur in these are not merely discursive but rather include a series of symbolic markers either from performances or local histories. Through these communicative characters, advocacy networks navigate their pluralities in a different way than do institutional forms of political processes. In creating multilevel understandings of purposes and activities, which at the same time relate to each participating group's agendas and priorities as to other actors' expectations, advocacy networks carry out ceremonies of civil becomings. In other words, advocacy networks' participants construct their own legitimacy as credible civil society actors as they advance in their negotiations to identify problems, state clearly their positions, negotiate common decisions, and announce agreements.

This process, as we have seen, is anything but smooth. The need for constant negotiation with positions that are markedly different from their own is partially why radical activists consider advocacy networks as too "soft." The more radical groups tend to prefer "hard" or "uncivic" forms of stating their discontent or dissent (Alvarez et al. 2017). And yet, in collaborating with reformist groups, radicals tacitly accept different ways to handle claims. While established political institutions, as local or federal governments, have their own customary forms of negotiations (sometimes by norms and agreed-on behavior, other times by de facto power struggles, or also by laws and institutional design), advocacy networks tend to be more flexible in how they deal with differences. Juris offered a view on this that he called the "cultural logic of networking": "Networking logics specifically entail an embedded and embodied set of social and cultural dispositions that orient actors toward (1) the building of horizontal ties and connections among diverse autonomous elements, (2) the free and open circulation of information, (3) collaboration through decentralized coordination and consensus-based decision-making, and (4) self-directed networking" (Juris 2008a, 11). This view, however, appears to situate as facts aspects I believe are intentions. People who decide to assemble first in groups or collectives, and then in larger networks are clearly seeking more than just the resulting collaboration. They are seeking increased influence in numbers and in the combination of knowledges and capacities. There is thus a conundrum regarding networks, as they are conceived as both structural arrangements and sociocultural configurations. The fact that as social assemblages networks can make decisions and act poses the question if they have agency. That is what the next chapter is about.

4

Entangled Agency in Networked Activism

Reality is complex, and I think that it is often convenient for people to be in doubt about precisely where IMA stands because there are very different and conflicting interest groups and stakeholders involved, and there is no reason why one needs to burn oneself with one group just to satisfy another. And I think it's honest, we do have differences, they're legitimate and we can handle it.

—Tony, senior researcher at IMA

IN NOVEMBER 2004, I took part in a two-day workshop organized by the Instituto do Meio Ambiente da Amazônia (IMA) and the Instituto Socioambiental (ISA). The workshop took place in an upscale hotel in Alter do Chão, an idyllic tourist spot that is only a forty-five-minute drive from Santarém. The purpose of the workshop was to establish a network of social leaders who were interested in reducing the negative impacts of the planned paving of the BR-163 highway, which connects the southern Amazon region in the state of Mato Grosso to the port of Santarém, in the state of Pará. Because the road crosses a densely forested area, the main fear was that rapid illegal deforestation would follow its paving. The workshop's participants included representatives of small farmers' unions, of small and big nongovernmental organizations (NGOs), and lawyers, community organizers, and even a small-town priest. All participants agreed that their position was not against the paving of the road per se, but rather against the government doing it without a plan to stop the illegal logging that usually follows the paving of roads in the Amazon. In the last meeting of the workshop, after an intensive two days of presentations and discussions, all participants needed to decide together what would be their first joint action. From the outset of the discussions, a clear tension arose between two camps: one supporting diplomacy, and another promoting radical dissent. "We will go to Brasília with a unified voice and make our case," said Silvia, who was acting as moderator of the closing meeting. A hum of disagreement arose, and Inácio spoke with visible agitation: "We need to show strength to be able to negotiate: we should make a show of force by blocking the BR-163 at the height of the harvesting season so that the soy will not reach the port to be exported. This way, they will surely pay attention and take us seriously." Some heads nodded in agreement. "I think that we have such a strong case that we do not need such measures," said Adrienne, supporting Silvia. A heated debate followed between a majority, who were seeking moderate action, and others who were pushing for a more radical response. Curiously, Silvia, Adrienne, and Inácio—the loudest voices of both camps—all worked for IMA. What was taking place in this gathering is an example of what I refer to as an emerging entangled agency.

In this chapter I analyze the concept of entangled agency as a form of agency that advocacy networks enact. In their process of civil becoming, advocacy networks not only use communication to establish and nurture the links that hold them together and the markers that characterize them, but they also flex their decision-action muscle through a series of practices that involves its social and material makeup. As seen in the previous chapter, the legitimacy that is therefore built throughout the life span of advocacy networks relies heavily on how decisions are made. To shed light on such entangled agency, this chapter explores two key aspects of it: what I call *netmentality*, or the increasing practice of linking up efforts among activists and advocates through a conviction that such networking is not only beneficial but also a requirement for sound work, and the *agency of networks*, as a distinct form of agency that is compared to other theorizations on the concept. Before starting with these topics, however, the following pages will develop the entanglements that make up a networked ethos, which I believe frame discussions of agency as something different from the aggregation of network members' capacities. These reflections are based on my fieldwork in Brazil.

In the meeting I already started describing, activists negotiated their joint intentions not only considering their own interests or agendas but also those of other stakeholders as well as of a series of nonhuman forms, such as the BR-163 road, the maps presented, and the produce from small-scale farmers and large-scale agribusinesses. The posters by which the meeting was promoted brought all these elements into play by showing the BR-163 road on a map of the Amazon with pictures of forest trees, soy production, indigenous peoples, and small farms, with the title of the event. The paving of the BR-163 highway was not opposed by those attending the workshop because it would benefit all stakeholders, and not just large-scale agribusinesses. For small farmers and inhabitants of extractive reserves, for example, it would allow more efficient distribution of their produce, as opposed to the difficulties they faced with the dirt road. The risk many of those attending the workshop insisted on was that illegal loggers could move very swiftly into new areas once the road was paved, as had already happened in other cases in the Amazon. This meant that decisions were not to be taken simply as oppositional. The proposal to block the BR-163, for example, would have a chain effect by denying the flow of soy to the port of Santarém and farther on to its final destination. Discussions about deforestation, however, constantly referred to the reduction in rainfall that was felt throughout the region and had been explained as a direct result of the increased forest clearings. In every possibility of joint action, therefore, members would consider and debate consequences for forest trees, animals, ecosystems, produce, and the various human communities that were to be affected by the paving. Such a level of entanglement, in this case very much informed by an ecosystemic approach (Pecqueur and Freire Vieira 2015), is part and parcel of this type of coalition. In a way, the participants were also already entangled in webs of personal and professional relations during years of activism and work in the region. Many of the participants already knew each other or about each other. They had gathered out of a shared need to devise a common strategy to face the government's decision to pave the BR-163 road. This highway had been originally planned by the military dictatorship in the 1970s as part of its scheme to colonize the Amazon region, but it was left unfinished, as a dirt road. Nevertheless, it was useful for traveling between the scarcely populated townships and farms that appeared over the years. With the exponential increase in soy production in the southern Brazilian Amazon, the road was used to transport the grain from Mato Grosso to Santarém, where it was loaded onto ships for its export markets. The fact that it was a dirt road made such transport slow and difficult, with only a minor part of the production going on that route and the rest heading to ports on Brazil's southern Atlantic coast. At the time of my fieldwork, numerous environmental collectives were exploring how to respond to the government's plans to pave it. The conveners of the workshop were IMA and ISA, two influential NGOs that are part of the socioenvironmental movement in Brazil, which promotes a balance between people and environment (as described in chapter 1). Their members were keen to strike an equilibrium between all stakeholders and the forest.

The agreement that all participants had reached to not oppose the plans to pave the road was due to the fact that everyone involved would benefit from the paving in one way or another. Small farmers would use the road to market their produce, and local populations could make some long trips in a shorter time. The risk they all acknowledged was that if the paving were to take place too quickly, without a plan to mitigate its potential negative effects, rapid deforestation would follow. The IMA had carried out long-term studies in other areas of the Amazon rainforest, showing how paving without planning for oversight and governance resulted in an exponential increase in illegal logging. As the disagreement portrayed above showed, however, negotiations among network members were often tense. Such tensions allowed for dissenting voices to be heard before individuals would work to seek to appease those who were upset and bring everyone into agreement. This process enhanced the network's legitimacy to present itself as a valid voice for making demands of the federal Brazilian government. Members of IMA and ISA had clear leadership roles, although they insisted they did not control the network. I agree with Keck and Sikkink, when they argue that "the agency of a network usually cannot be reduced to the agency even of its leading members" (Keck and Sikkink 1998, 216), which means that it is not possible to identify a unified source of decision and action, or agency.

Recent work on distributed agency has yielded in-depth analyses of humanity's entanglements with objects and nonhuman life forms (Enfield and Kockelman 2017). Entanglements are considered in anthropology as the crux of what any thing or any person is. Ingold has led the way in arguing for the importance of paying attention to how all life forms and objects are entangled as well as for avoiding simplifying the relations that exist between them (Ingold 2008). Regarding agency, Alfred Gell has argued that artifacts are already so enmeshed in a texture of social relationships that they can be considered as agents in themselves (Gell 1998, 17). Agency is a concept that refers to individuals' capacities to make decisions and act (Emirbayer and Mische 1998). This conception is already laden with an assumption that humans *are* free to decide and act, and that such freedom has been fundamental to the shaping of history. As Gell describes, "Agency is attributable to those persons (and things . . .) who/which are seen as initiating causal sequences of a particular type, that is, events caused by acts of mind or will or intention, rather than the mere concatenation of physical events" (Gell 1998, 16). In many cases, critical perspectives have nevertheless tied the concept of agency to that of structure, which refers to the symbolic framework that may well constrain people's potential decisions and actions. The debates on these concepts have helped shape scholarship on power and knowledge. They create a forum for exploring the way certain forms of knowledge are legitimized by power structures, and how such forms of knowledge may help maintain the status quo of the power structures that legitimized them. These considerations of circular flows of legitimation are not meant to convey a static world in which there is a single source of power that constrains other forms of expression. They rather shed light on negotiations and conflicts through which social collectives define their symbolic and material paths. Agency is therefore a useful looking glass for studying activism and advocacy, as these consist of resolutions and deeds that commonly challenge established norms of sociality (Kennelly 2009; Atwal 2009). In disciplines such as sociology or political science, the study of advocacy and activism is usually focused on organizations as actors in institutional settings (Tarrow 1998). This fits in with a macroview of human history as defined by competing power clusters in the form of governments, "big men," or enterprises (Tilly 1978). In anthropology, the preferred route is to scrutinize the social interactions within and among groups that may illuminate human diversity and also provide clues for better understanding large-scale dynamics.

By deliberately forming what they themselves call "networks," activist groups and nongovernmental organizations expose their attitude toward the collaboration. Network members have made key claims about their networks that distinguish them from other forms of association, such as: (a) *decenteredness*, (b) *equal inclusiveness*, and (c) *organizational flexibility*. All of these should be considered as *claims* by those who form the networks with the purpose of describing them and defining their scopes and functions. *Decenteredness* is a principle often set against "vertical" organizational arrangements, where one or a few influential groups make decisions for the whole. In claiming a web structure, some groups argue that vertical arrangements are nonexistent. Such a view seems to be a mantra among activists—inspired by anarchist principles—arguing that no group within a network should exercise power over others (Maeckelbergh 2009; Juris 2008a; Graeber 2009). In my observations, I found several cases where this was more a desire than an actual practice. It is common for groups with more experience, resources, or legitimacy to exercise strong influence over other network members. The difference is that such influence is no longer embedded in the organizational chart and is exercised openly in a centralized manner. Regarding *equal inclusiveness*, networks claim to be constituted by the plurality of their members, arguing that as such makeup changes, so does the whole network. This principle is also posed against a centralized conceptualization of what a collective is. Again, this is usually more a wish than a reality, which often nevertheless serves as a horizon, as a goal to reach, or as a moral compass. The way activists described this characteristic to me is that while in other structures a group joins a collective that is already set, in a network each new member may influence what the network is. In terms of *organizational flexibility*, activists seem keen to stress that networks are a refreshing arrangement that are nothing like long-established bureaucracies whose set ways result in delayed decisions and changes. Although this seemed to be the case with most of the networks I observed, there were also some signs that if the network were to exist for longer, it would need to develop bureaucratic processes that would allow for legitimacy in decision-making among its members.

The use of the "network" as a guiding organizational principle by activists and advocates is thus meant to signal a certain type of system that aligns with its members' high hopes for effective and fair collaboration. Riles's reflection on advocacy networks in the Pacific is relevant here: "The Network's claim to spontaneous, collective, and internally generated expansion and its ability to create systems that preserve the heterogeneous quality of their elements imbues its extension and enhancement with a certain normativity. Its existence is a good in itself. No one, it would seem, could possibly be 'against' networks (whether or not they achieve other ends), for the Network is simply a technical device for doing what one is already doing, only in a more efficient, principled, and sophisticated way" (Riles 2000, 173–74). Activist or advocacy networks with apparently clear aims need to make decisions regarding joint actions, demands, and resolutions. Such decisions must have a certain level of legitimacy for the networks' members to feel included and represented in the collective efforts. This process is evident in all scales of engagement, from transnational to local.

During my time in Brazil, I witnessed similar scenes to the one portrayed at the start of this chapter in the various networks of which IMA was a part. The workshop referred to, which took place in a hotel in Alter do Chão along the Tapajos River, brought together people with many different levels of expertise who carried out activities in smaller groups in an attempt to stop the rapidly deteriorating environmental balance in their corners of the Amazon forest. Several of those present told me separately that they were in the meeting because they wanted to collaborate on a joint project. "We are not here to follow anyone, but to give shape to our own idea of what the Amazon should be," one environmental activist told me. All activists I spoke to in Brazil and Barcelona considered themselves as guarantors of a search for justice that would not take place without their intervention. Each of them considered their place as a consequential part of a larger whole. Such a notion is thus a case of emergent properties arising out of the combination of its constitutive elements (O'Connor 1994). The concept of *assemblage* is here relevant, as the groupings that are gathered in networks labor to alter complex aggregations of things and people. DeLanda clarifies Deleuze's view of assemblages, defining them as "wholes whose properties emerge from the interactions between parts" (DeLanda 2006, 5); he adds that those same "properties may be irreducible to its parts but that does not make them transcendent, since they would cease to exist if the parts stopped interacting with one another" (DeLanda 2010a, 12). For example, the way one part interacts with others is not clear when seen by itself; it is the interaction that elucidates the role of each part. Also, "a component part of an assemblage may be detached from it and plugged into a different assemblage in which its interactions are different" (DeLanda 2006, 10). Marcus and Saka explore the use of assemblage for analyses of social processes as a concept that "can refer to a subjective state of cognition and experience of society and culture . . . or it can refer to objective relations" (Marcus and Saka 2006, 102). This is why the term *network* lends itself not only to the collaborative setup that activists use but also to their particular take on such collaboration. Individuals participating in such networks constantly remind one another that because they form a network together, their joint actions *should* take a different path, both from their group path and from the paths of other associations.

Netmentality: Established Collaborative Ethos

"Dozens of NGOs are working to reduce fires in the Amazon . . . but what good is it if each one works in a different corner of the forest and does not collaborate with others?" an activist asked me rhetorically, during a coffee break at a workshop to help farmers avoid accidental fires. She was explaining the benefits of participating in networks. The small NGO she had founded a few years before had survived thanks to funding from large foundations, which in turn had asked her to take part in collaborative networks with other organizations. She said these collaborations had provided learning experiences that allowed her and her staff to grow. I heard similar opinions from many activists and NGO personnel in different meetings and sites around the Amazon. When I attended a large conference of the United States Agency for International Development (USAID) in Manaus, I understood at least part of what motivated similar reflections. The conference consisted of seminars and talks quite similar to academic ones, but also included meetings of thematic networks funded by USAID. "We are interested in creating synergies between active groups," an official from the development aid agency told me when I asked him about the purpose of such meetings. Each NGO that received funds from the agency was required to form part of at least one thematic network. The networks with the largest membership were those aiming to reduce forest fires, to stop illegal logging, and to promote extractive reserves. The same official explained to me that the networking logic that was required of the participating groups was a way for them to avoid competing against one another for funds and to promote collaboration.

While these networks were of NGOs and other activist groupings, national aid agencies and foundations had been instrumental in their formation. By promoting the development of more networks, funding bodies thus established such organizational arrangements as a desired form of collaboration.

It is a case of what I call "netmentality," paraphrasing Foucault's governmentality. For Foucault, it is not possible to study technologies of power without an analysis of the political rationality behind them (Foucault 2008). Although the concept of network does not imply a similar overarching form of social control or oversight as that of government, I suggest that it nonetheless provides a model for visualizing and ordering interactions and thinking about causalities. Just as professional politicians seek to embed the population with a way of thinking that implies a central role of the state and its institutions for maintaining social order, so too for today's advocates is it essential to promote methods of networked collaboration that can result in individuals feeling an agentic sense of proprietorship of the collective, that is, that can make polity members feel they are in control of a polity. It is what Foucault called a *dispositif*, or the "the material, social, affective, and cognitive mechanisms active in the production of subjectivity" (Hardt and Negri 2009, x). The network is thus not here considered simply as an analytic concept through which we can interpret human relations, as was originally introduced with social network analysis (Barnes 1954; 1969), but rather as a semi-institutionalized form of engagement. Through technological innovation, various layers of physical and virtual interconnection have expanded our awareness of our relations to others and of multiple and complex causal action (Latour 2013, 33). Urban layouts, sewage, pipelines, roads, railways, telephone lines, and electromagnetic waves are examples of how technological networks have increasingly become the existential infrastructures of our times (Mattelart 2000). The cultural impact of such interconnection is considerable, as it shapes an emerging imagination of what is possible (Van Dijk 1999). Most indigenous groups in the Amazon, for example, are not only in contact with government authorities of their respective states but also in relationship with other groups for various reasons. Some ethnic groups even have their own NGOs, which they set up with the help of established NGOs that in turn channeled funds from international foundations and aid agencies. In this way, indigenous communities are not only beneficiaries of others' efforts, but they are also active participants in advocacy networks. This raises the question: are they being agents or responding to structures?

In her ethnography of a web of groups working toward the 1995 UN Conference on Women in Beijing, Riles stressed the way in which the network reified itself (Riles 2000). Among the members of participating NGOs, the acts of "keeping in touch," "reaching out," or "soliciting your views" were ways of shaping the networks, which Riles defined as "institutionalized associations devoted to information sharing" (Riles 2000, 59). Riles was particularly mesmerized by the aesthetic value of the documents that were produced as part of the negotiations and ongoing preparation of the network of NGOs.

She compared the documents with Fijian mats: "Both were collective, anonymous, and highly labor intensive exercises that required great attention to detail. Both kinds of production ultimately yielded objects collectively acknowledged as highly valuable and a source of pride to their makers. Like mats in Fijian ceremonial life, moreover, the documents provided the concrete form in which collectivities (whether groups of clans, persons, or organizations) were 'taken to' another environment" (Riles 2000, 73). Riles's analysis characterizes networked arrangements as structuring structures, that is, as configurations that shape the interactions that occur within them.

This approach paints a picture of networks as renewed forms of bureaucracy. The ones I observed, however, actually worked differently from this image by balancing attempts to achieve predictable bureaucratic decision-making processes with efforts to maintain a certain degree of anarchic randomness and openness. I consider this combination the strength of such advocacy networks, as they avoid stalemates resulting from overinstitutionalization while balancing quite different ways of doing things. As an organizational principle, the network was identified in the twentieth century as a model to improve the management of information flows. This was especially useful to, and therefore quickly applied in, industry, commerce, and government. It is at the core of what several authors have named the "network society" (Castells 2010; Van Dijk 1999). Although corporations and governments did not openly name their rearranged structures "networks," in practice they reorganized chains of supply, distribution, and trade through a distributed web of nodes that avoided centralization (McCarthy, Miller, and Skidmore 2004; Mendizabal 2006). For Annelise Riles, exchanges and interactions within networks that reach across localities toward what is known as the global are better understood by their "informational aesthetics" (Riles 2000, 20). Her proposition coincides with Foucault's interpretation of the power/knowledge regime (Bevir 1999). Through his concepts of biopower and governmentality, Foucault explored forms of surveillance and ideological control through which modern institutions have established the primacy of the state as a political unit of societies around the world. In his perspective, regimes of power/knowledge determine the standards by which individuals should behave, with disciplinary and punitive technologies to ensure compliance (Foucault 1991). These reflections, however, appeared to leave individuals with little option but to form themselves new regimes of power/knowledge in order to challenge the existing ones.

In some scholarly reflections on networks, there appears to be a romanticizing of their potential. Riles points this out by stating that "in international law and international relations theory, for example, networks as observable institutional organizations of governments and NGOs are widely viewed as more flexible, more progressive, more sophisticated forms of international action, which hold out the hope of success where the state system has failed" (Riles 2000, 172). This positive spin has convinced foundations, aid agencies, and other donors to become driving forces for the creation of networks, as the ones portrayed here. Some foundations and international NGOs provided texts to activists and local NGOs explaining the potential benefits that networks represented. The Brazilian chapter of one large international NGO, the World Wildlife Fund for Nature (WWF), for example, produced and distributed a book titled *Networks: An Introduction to Dynamics of Connectivity and of Self-Organization* (Costa et al. 2003). After I visited Chico Mendes's house in Xapuri, in the state of Acre, a friend who works in the Rio Branco offices of the WWF Brazil gave me a copy. The book's cover has a photograph of a beehive, which is a common symbol of collaborative efforts. Other key actors of the socioenvironmental movement have continued using the concept of network as central to their proposals for political change in Brazil. Above all, Marina Silva, who grew up in a rubber tapper family, joined the Workers' Party from a young age, was minister for the environment under Inácio "Lula" da Silva, and became one of the most visible defenders of the Amazon and the socioenvironmental agenda, named her new political party Rede Sustentabilidade (Sustainability Network). These are just two examples of the reach of the concept that is used as a mold for collective action. Effective networks, however, require a clear link to a locality and its issues. As Routledge argues, "Successful international alliances have to negotiate between action that is deeply embedded in place, i.e. local experiences, social relations and power conditions, and action that facilitates more transnational coalitions" (Routledge 2003, 337).

So what happens to the individual in a world where networks are performed as structuring structures? If subjectivities are formed through quotidian interactions that are themselves known as "networked," then the will to take part in them is not entirely born out of a creative mind but rather out of an already shaped one that identifies networks as important organizational structures. The cultural implications of networked milieus around us thus also challenge the conception of political agency (Mulgan 1997). Social assemblages that seek to influence governments or corporations that are already networked use the same network logic to be able to compete with the organizational sophistication they face. What role does agency play in this setup?

To come back to the example of indigenous NGOs, I asked Sergio from São Paulo while we were taking a break from a meeting, "Would these communities have developed their own NGO without external help?" His job at the influential ISA was to support the ongoing work of ATIX, the NGO of indigenous communities in the Xingu National Park that ISA helped set up. "Perhaps not," he answered, "but through it they can really decide for themselves what projects they want to carry out and seek funding for." This conversation unfolded as we walked in Canarana, a city that was only founded in the 1970s in what was back then a densely forested area of the state of Mato Grosso. Today, the city is surrounded by large soy plantations, and the Xingu National Park to its north is the only remaining large forested area. We were attending the launching event of a campaign to protect the springs that feed the Xingu River, which saw for the first time representatives of the Grupo Maggi, the single largest soy producer in the world at the time of fieldwork, sitting down to negotiate agreements with small-scale farmers and indigenous populations, among other participating groups. The campaign, called Y'Ikatu Xingu, sought a purposely complex approach to a difficult situation: even though the Xingu River and a large portion of its basin remained protected within the Xingu National Park, many of the springs essential for providing water to it remained isolated within soy plantations, where encroaching farming threatened to dry them up, and whose chemicals used as pesticides and fertilizers had found their way into their streams, polluting the whole basin and river. The campaign lasted ten years, from 2004 to 2014, and mobilized numerous efforts to raise awareness of the problem and change governmental policies. A new, similar large-scale campaign was launched afterward to collect, protect, and share seeds from local plants, and to promote more public knowledge about them in an effort to restore regional vegetation.

Environmental advocacy requires complex approaches because problems related to the health of the planet escape simple causal relations. Networks and campaigns, as the ones portrayed herein, therefore, may bring together actors (both individuals and groups) who would otherwise not collaborate. This type of network, linking not only NGOs and activists but also businesses, schools, academics, and others, is thus quite different from those portrayed recently as part of the "global justice movement" (Juris 2008a; Graeber 2009; Maeckelbergh 2009).

Not all advocacy networks, however, operate in the same manner or prioritize the same types of objectives. In order to attempt a first classification of contrasting approaches, I distinguish two major types of networks as "dissonant" and "harmonious." Both emerge from the will of their members, usually in the context of an issue all find important. The difference is that while the former purposely seek the participation of groups with contrasting positions as a necessary tactic, the latter prioritize a shared understanding that comes to define the network. One is dissonant because not all members would normally identify with one another, but they collaborate out of a shared sense of moral obligation; the other is harmonious due to an overarching identification among participants despite their differences, usually in ideological terms. One of the networks I witnessed in the Amazon is a good example of a dissonant type. It was the Y'Ikatu Xingu campaign I mentioned above, which brought together numerous actors with the purpose of collectively tackling a single problem that, to varying degrees, affected all the participants: the health of the springs that feed the Xingu River's tributaries in the state of Mato Grosso. More specifically, the shared aim was to improve the quality of the water reaching the Xingu River (one of the tributaries of the Amazon) while also improving ecosystems in the area. Officially called a campaign convened and coordinated by ISA, it worked as a network for ten years combining practices by dozens of very different groups, from schoolchildren to indigenous collectives. Harmonious types, by contrast, may include some alterglobalization networks, as their members join efforts because they all share a strong set of beliefs (for example, that corporations are leading an unhealthy type of development). In my assessment, both network types share a similar view on the underpinning reasons for problems. The main difference is that one invites corporations to the negotiating table as stakeholders, while the other prefers to shame them publicly as culprits. Perhaps the difference lies in a pragmatic approach in one (due to its internal plurality) and an ideological commitment in the other (due to its cohesion).

Several good ethnographies of activist networks have been published over the last few years, but Juris's *Networking Futures* stands out because of its analytic depth and reflective narrative (Juris 2008a). It is an account of Juris's immersion in global justice movement networks in Barcelona and around the world. In his multisited ethnography following various events and protests, he describes the shared ethos of seeking an alternative form of governing life and practical forms of influencing government policies to ensure those views. He explains that "anti-corporate globalization movements involve an increasing confluence among network technologies, organizational forms, and political norms, mediated by concrete networking practices and micropolitical struggles" (Juris 2008a, 2). His focus on protests, marches, and other public forms of demonstrating the activists' dissenting views, however, highlights how he carefully stresses collaboration among radical groups and downplays the role of NGOs (Juris 2008a, 40). Activists often criticize NGOs and moderate groups, Juris goes on to explain, because of their "reformist" agenda and their "top-down" organizing practices. The network is therefore considered by Juris and other researchers of radical activism to be the alternative organizing principle to a perceived authoritarian tendency among many established groups, such as NGOs. Academic stress on direct action as an array of public displays of dissent by activists is therefore often characterized as an alternative to traditional political arrangements. Examples are Rheingold's smart mobs (Rheingold 2002) and Hardt and Negri's multitude (Hardt and Negri 2005). Although they are interesting, I cannot help but consider these analyses of networked political collaboration as too narrow; they assume that people already share a substantial worldview.

Increasing anthropological attention to NGOs, on the other hand, has yielded fascinating insights into the tensions in which these organizations exist. As important actors of what is known as the "development industry," NGOs have been commonly criticized by scholars for their role in depoliticizing situations because of their involvement with aid agencies that seek technocratic responses to certain problems (Ferguson 1994). Critical studies of development have thus turned a skeptical eye on NGOs, highlighting the distance that often prevails between the good intentions of their members and the structural limits of what they can attain (Schuller 2012; Beck 2017; Fisher 1997; Choudry and Kapoor 2013; Hulme and Edwards 1997). But some have also reflected on their organizational flexibility and adaptability in light of problems or doubts (Lewis 2014). The fact that NGOs can access resources through formal applications and institutional representation and apply them to projects to directly address problems makes them important political entrepreneurs. Regarding their transnational collaborative potentials, Schuller has argued that NGOs complement neoliberal strategies at the global level by providing legitimacy for its agenda, undermining aspirations for social welfare state apparatuses worldwide, reproducing inequalities with their labor practices, and representing institutional barriers against local participation and priority setting (Schuller 2009, 85). Anthropological scholarship on NGOs has grown exponentially by exploring the vast diversity of settings, practices, and implications of the work of NGOs.[1] Furthermore, it is common for anthropologists to collaborate with NGOs, often through a contribution to the generation of knowledge of local customs, cultural translation, or mediation.

Advocacy networks, however, cannot be situated either as solely radical or simply as aggregations of NGOs. As collaborative events, they defy our understanding of a group. It is therefore relevant to inquire further about the way their agency works. And that is what the next part of this chapter tackles.

Agency of Networks: Collective Action in a Networked Fashion

The concept of agency has most commonly been used in regard to tension between an individual and the social collective. It is anchored in the

1. The evolution of recent anthropological studies of NGOs can be best understood in the special issues of *Political and Legal Anthropology Review* of 2001, vol. 24, no. 2, edited by Leve and Karim; and 2010, vol. 33, no. 2, edited by Mertz and Timmer. Schuller and Lewis have also provided a valuable genealogy of NGO scholarship in an Oxford bibliography (www.oxfordbibliographies.com).

centuries-long debates about free will. As Giddens notes, "Agency concerns events of which an individual is the perpetrator, in the sense that the individual could, at any phase in a given sequence of conduct, have acted differently" (Giddens 1984, 9). Having agency therefore implies both the decision and the ability to act: "An agent is one who 'causes events to happen' in their vicinity" (Gell 1998, 16). One of the most common questions is whether people have the capacity to make their own choices and act on them in relation to the structural configurations under which they live. The main tension is between the degree to which individuals are able to make their own decisions in spite of structures and the extent to which their decisions actually allow the social structure to continue. A resulting paradox is that while "the individual has agency in the sense that he or she is a creator of society[, s]ociety imposes structure on the individual and limits his or her options" (Eriksen and Nielsen 2013, 54). Human behavior, therefore, is a sociocultural negotiation between the individual and her social milieu. In anthropology, such a social milieu is understood as a system made up of relations—often forms of identification, such as kinship or ethnicity—that maintain social life. Most attempts to analyze agency, however, "end up privileging one or the other of agency's antinomies, appearing either too voluntarist or overly determined by structure" (Patterson 2006, 211). Herzfeld, however, argues that there is actually more of a balance: "social life consists of processes of reification and essentialism as well as challenges to these processes" (Herzfeld 1997, 26).

Agency is here also considered a capacity of a social assemblage, either in the form of a group or of a network. Social collectives do become agents when they make decisions together and act on them. Such processes are of a different order from those of individuals. List and Pettit have recently put forward a philosophical theory to substantiate group agency (2011). They draw on authorization theory and on what they term "animation theory." Authorization theory, they explain, was first postulated by Hobbes to analyze the way in which a collective authorizes the following: an individual to speak on the collective's behalf; the form that collective organization takes; or a small group to act with the collective's consent (List and Pettit 2011, 7). Animation theory, however, stems from an organicist view of emerging group behavior at a biological level (List and Pettit 2011, 9). List and Pettit go on to clarify that a collection of individuals jointly intend to promote a particular goal if four conditions are met: if they have a shared goal, if each participant intends to carry out a particular action for the common goal, if there is interdependence among members, and if there is a common awareness among all involved about the three previous aims (List and Pettit 2011, 33). The authors develop their idea of a group agent as a rational entity that is actually controlled by each individual participant's will to associate (List and Pettit

2011, 160). List and Pettit's main focus, however, is on the performative: they claim that what makes an agent "is not what the agent is but what the agent can do" (List and Pettit 2011, 176). It is worth remembering, however, that for anthropologists "'folk' notions of agency, extracted from everyday practices and discursive forms, are of concern" over those considered "philosophically defensible" (Gell 1998, 17).

List and Pettit's level of abstraction is nevertheless useful for thinking about advocacy networks, as it is possible to argue that in their connected state its members are able to conjure up a collective intentionality and action. But here I need to insist on the differences between *group* and *network*, as both were clearly distinct forms of collective in my fieldwork. Whereas a group aspires to have a common identity and work with a unified strategy, a network seeks to negotiate its purpose usually as a temporary effort toward a goal. "Each NGO has a niche or specialty, and we collaborate when there are problems that involve several of us," explained Renata, who was the executive director of IMA during my fieldwork. NGOs in the Amazon increasingly collaborate in networks. While both groups and networks are "plural subjects" (Gilbert 2006) and are characterized by unified intentions (Bratman 1999), the group is more coherent than the network, perhaps due to a type of formality of bonds that holds its members together. In networks, each participating associate is not bound to remain, instead choosing to do so out of a conviction of its benefits. While one could say that an individual within an NGO could choose not to be part of that NGO, the group would not change because of his leaving. In networks, so goes the mantra I heard repeatedly from activists, each participating member influences the whole. One common way of describing the difference is by contrasting vertical versus horizontal forms of distributed decision-making and action. Castells explains that corporations went through such change from vertical to horizontal forms of power management in the twentieth century "to adapt to the conditions of unpredictability ushered in by rapid economic and technological change" (Castells 2010, 176). The vertical refers to an executive body made up of a minority of members—either one person or a team—that makes decisions for the group. The horizontal refers to a less hierarchical form of decision-making—one that requires negotiation among members in order to reach a decision. These forms, however, are never "pure": the vertical always implies attention paid to opinions or agendas of other group members, and the horizontal usually involves various forms of leadership. Group agency is marked by a capacity for joint representations, motivations, and action. "Thus the group is organized so as to seek the realization of certain motivations in the world and to do so on the basis of certain representations about what that world is like. When action is taken in the group's name—say, by its members or deputies—this is done for the satisfaction of the group's desires, and according to the group's beliefs" (List and Pettit 2011, 32). Because of the network's logic of loose connections, it is perhaps more convenient to consider its form of agency as an outcome of its members' entanglement.

The IMA is an NGO dedicated to carrying out scientific research on the Amazon environment and using the results to advocate for better policies for the region. Its scientific staff is made up of academic researchers who teach at universities in Brazil and the United States. They were thus curious about my project, about which we corresponded for six months, before they granted me access to their offices and projects. After my arrival at IMA's headquarters in Belém, I was gradually introduced to key personnel in charge of the NGO's different projects. It took over three weeks to meet the basic team, not because it's so large, but because they are constantly traveling to research sites or to meet stakeholders, other environmental groups, or policy makers. I knew that a lot of what they did was dedicated to networking—building and maintaining connections with other groups, such as environmental NGOs, funding bodies, or with policy makers—but I had not expected it to be so central to their work. Each of the sections within IMA was dedicated to a project, and each would be IMA's representative in one or several networks relevant to that project. This arrangement was quite similar to academic organizations, as each section applied for project grants from funding bodies and grew or shrank according to its success at receiving them. Some areas within IMA were not very good at attaining large endowments from abroad but were helped by the NGO's general council because they were considered important to the achievement of its overall aims. Such was the case of a team dealing with community organizations, dedicated to grassroots political organization that would build up concerted efforts against various of IMA's key interests (such as deforestation or reduction of fire). This team was quite critical of what they saw as the NGO's neoliberal agenda. Its members tended to side with left-wing activists in larger meetings (as happened in Alter do Chão).

The division within IMA illustrated a widespread demarcation present in civil society organizations (NGOs and social movements) between individuals following agendas with radical and reformer objectives. Members of both camps seek a change in how the Amazon region is managed, but their opinions on how this should be achieved vary—sometimes quite starkly. Within IMA, the largest group by far would be on the side of the reformers. Most of these individuals are scientists who are interested in understanding various aspects of the Amazon ecosystems. One example is Tony, an American scientist who has studied political ecology in the Brazilian Amazon for thirty years, who is married to a Brazilian, and who lives in Belém. He is one of

IMA's leading scientists. His focus on smallholder settlements combining conservation with resource management has proved quite successful in the design and implementation of floodplain fisheries, among other projects. As we were having coffee one day on the terrace of his house in Belém, he insisted on the need for dialogue between people who hold different opinions. However, he argued that there was a need to keep the most radicalized voices out: "If you have a polarized debate the radical elements dominate and the interests of most people are lost. So the aim must be to separate the reasonable people from the unreasonable people and have that core expand. So I see that our approach has got to be to help facilitate that dialogue and interaction."

The smaller group of radicals at IMA, however, would often disagree with the institutional dealings the NGO maintained with large agribusinesses (mainly soy producers). Mariana openly identified herself as a radical. Her job within IMA was to serve as a political liaison between IMA and local community organizations (mostly small farmers' unions) in order to establish and maintain working relations with the aim of successfully carrying out a number of IMA research projects. Being in constant contact with local activists who faced violence from illegal loggers or miners made her distrustful of what she perceived to be the reformers' patience. On several occasions, I heard her disparage her bosses and other researchers at IMA. I traveled with her to visit several communities and attend a few workshops with local leaders. On one occasion, after a whole-day meeting with small-scale farmers who had traveled nine hours on dirt roads, Mariana complained to me that IMA's strategy was following the interests of big capital. She also complained that the high-profile bureaucrats who attended the meeting were out of touch with the region, as we had seen them in the hotel restaurant complaining about the lack of food choices and the poor quality of the hotel. The whole-day meeting we had just attended had largely consisted of official speeches and many promises, with a couple of practical sessions to help farmers understand new government policies. That night, while having an after-dinner beer, she told me that "we need a different approach to help these farmers live dignified lives. Sometimes I feel we're just playing the game of hide and seek with big soy producers." From her tone and demeanor, I got the sense that she wanted out. She appeared frustrated and somewhat depressed. A few months later, however, I asked her again how she felt about the differences between her way of understanding power relations and those of the reformers. The conflict that arose at the workshop described at the start of this chapter had just taken place, which is why I raised the issue directly at this juncture. She took a breath before calmly saying, "We do not agree on how to do things generally, you know? But they [IMA's reformist leadership] appreciate a dialogue with us, and I think that is important." Her position confirmed my suspicion that IMA actually finds the tension between both positions to be a productive asset. I thus consider IMA an arena where plural opinions are encouraged. Tony acknowledged the differences within the organization and argued that they were fundamental to IMA's success: "I think we have done a very good job of having good trustworthy relationships, and of administering areas of disagreement and competition for space reasonably well. I think we have very adept political negotiators in people like [Silvia, and Adrienne, and Mariana], who may not always agree with each other, but who share a fairly common vision of where they want to go and how they can get there. And potential conflicts get resolved or at least don't erupt in threatening situations. I sense that tension is actually very productive, because it obliges them to have a dialogue, and in the end I can see that it makes it more effective when taking into account other points of view."

The networks in which IMA was active were dissonant, that is, they had little internal cohesion. This meant that their members often needed to deal with contrasting opinions. It was an ongoing effort by some groups to try and maintain a unity that would enable trust and cooperation among the diverse groups in the network. A farmer who spoke at the meeting in Alter do Chão summarized the raison d'être of the network when he explained his own reason for becoming a leader against violent groups seeking profit in illegal logging and mining as being related to his belief that people are stronger when they are in a group: "If we stay alone, we cannot do much, but if we act together we may be able to stop them [illegal loggers]," he said, still trembling after having described how he and a friend had recently been shot at while walking home. His friend was killed, and he himself barely escaped unharmed. As he spoke, his voice cracked, and Rosa, a worker from ISA who was standing beside him, put a calming hand on his shoulder. The applauses that followed were accompanied by messages of solidarity: "We are with you, man." All those attending knew how dangerous it was to take a stand against violence-prone groups. It is a common tale in the Amazon frontier, where deforestation runs parallel to bloodshed (Branford and Glock 1985). Such violence has been an essential element in galvanizing efforts and gaining support for environmental networks. An example can be found in the aftermath of the assassinations of Chico Mendes and Dorothy Stang (Hochstetler and Keck 2007, 1).

Chico Mendes was foundational to the growth of networked efforts in the 1980s when he launched the "peoples of the forest" coalition, linking indigenous, rubber tapper, and environmental groups in a common cause (Mendes 1991). One key idea that came out of such combined efforts was that of the extractive reserves, which are conservation areas where a caretaker local community can carry out sustainable forms of resource extraction (e.g., hunting, fishing, or harvesting wild plants such as Brazil nuts) (Vadjunec and Rocheleau 2009; Hall 1997). This idea has been turned into government policy in Brazil. It is an example of entangled agency, as the whole ecosystem that would allow for the extraction of rubber tapping, Brazil nuts, acai, and medicinal plants and other life forms and minerals, was considered fundamental to any decision about the fate of the region. The "peoples of the forest" therefore operated as a network of activists, scholars (including two anthropologists), NGOs, and other groupings, which allowed for fertile discussions and ideas and managed to involve government actors. The fact that their proposals, like the extractive reserves, have become state policies proves a capacity to collaborate and persuade reluctant power holders of alternative solutions to identified problems. Like Hochstetler and Keck recognize, "The extractive reserve embodies a classic enabling strategy in that both state and societal actors need to collaborate continually to achieve the reserves' ends. State actors are needed to formally establish both the overarching legal framework for collectively held property and specific territorial reserves" (Hochstetler and Keck 2007, 162).

Latour's actor-network-theory (ANT) is relevant here, as the forest itself, its inhabitants, the many objects that are used in it, and the activities that threaten it (such as ranching or large agribusiness) are part and parcel of what is discussed and negotiated. For Latour, it is necessary to "follow the actors themselves" "to learn from them what the collective existence has become in their hands, which methods they have elaborated to make it fit together" (Latour 2005, 12). In the traces from ongoing interactions, Latour argues, one can find significant moments that define practices: "Action is not done under the full control of consciousness; action should rather be felt as a node, a knot, and a conglomerate of many surprising sets of agencies that have to be slowly disentangled" (Latour 2005, 44). Some analyses of NGOs and the networks they form tend to overemphasize their "modernizing" role. It is easy to see why. As somewhat bureaucratized organizations interested in establishing standardized forms of managing affairs (relating to social, environmental, or other issues), they often apply regulations attuned to market forces in an extended form of governmentality (Peterson 2001). But such reading runs the risk of simplifying interactions and stressing a linear causality that is not there. The networks NGOs form part of can be quite extensive and convoluted, which is why the ANT framework may be helpful to understand the flows that pervade such networked arrangements. As Latour notes, "Action is borrowed, distributed, suggested, influenced, dominated, betrayed, translated" (Latour 2005, 46).

In the Y'Ikatu Xingu campaign, local schools played a role by bringing students to various springs to study their water quality. When school officials are part of the campaign itself, making decisions and feeling included, their participation is more enthusiastic than it would be if they were to join something that began operations without them. The name Y'Ikatu Xingu, which means "good, healthy water" in the Tupi language, was selected by all those attending the launch of the network, by vote. ISA convened the network and used an inclusive language and methodology to make sure that all stakeholders involved would feel a sense of ownership. To close the launch ceremony in the town of Canarana, in Mato Grosso, a group of indigenous men performed a "war dance" to stress their nervousness and distrust after having felt constantly betrayed throughout their lives. During its ten years of existence (2004–2014), Y'Ikatu Xingu operated as a network with combined activities from all its members to contribute to its unifying aim of improving the quality of the springs that feed the Xingu river. Actions included planting trees, monitoring water quality, developing methods to restore degraded areas, training environmental educators, and spurring other networks, like the Xingu Seeds Network (Rede de Sementes do Xingu).

Socioenvironmental networks in Brazil are part of a global trend in the last few decades of collaboration between NGOs and social movements, in networks that have been credited by scholars and journalists with substantial influence at international and national levels (Keck and Sikkink 1998a). These organizational arrangements are increasingly acknowledged as powerful nonstate actors in a political arena traditionally dominated by state officials (Macdonald 2008). Analyses by scholars working on international relations or political negotiations tend to consider such networks as new members in an already crowded playing field. What such organizational assemblages represent is an innovative reconfiguration of the political. "The genius of NGO networks is in their ambiguity," argues DeMars (2005, 57), who considers these networks a "wild card in world politics." These networks make decisions and act on them with flexibility and a fragmented operative capacity. Their main internal challenge is to maintain a level of legitimacy in decision-making and action that holds the network together, as member groups or individuals can opt out at any moment. While attending meetings, workshops, and visiting projects of various networks in the Amazon, I noticed how activists and advocates paid attention to the ebbs and flows in the interests and motivations of individuals. Such attention was one ingredient among many that influenced the types of decisions that were made each day, decisions that ended up having an influence in public policies. Could this be understood as a reshaping of political power? Most of the activists I spoke to in Brazil like to think so. Like Brettel suggests, "Often, in fact, law and policy follow people; structure responds to agency" (Brettell 2002, 441).

This arrangement helps explain the concept of entangled agency. Whereas "group agency" implies a uniformity that provides the group with a type of personhood (List and Pettit 2011), "entangled agency" is not so much concerned with the network as an actor but with the particular moment and place where conditions and opportunities converge and open up the possibility for joint action. Although scholarly debates about "group agency" have considered various types of associations and informal groupings, it is common for groups themselves to be considered as units in order to address legal aspects and institutional entanglements. The essence of a group is the continuity of its relational arrangements: "Any multi-member agent must be identifiable over time by the way its beliefs and desires evolve. So there must be a basis for thinking of it as the same entity, even as its membership changes due to someone's departure or the addition of new members" (List and Pettit 2011, 32). Here, List and Pettit refer to a "multi-member agent," by which they mean a "group." Network agency may be better understood if defined as complex: "complex agency involves more than a connection to meaning, goals, and action; it also involves a relation between at least two beings with such powers" (Connolly 2011, 26). It is a type of "creative agency" that allows for novel forms of behavior, and that is facilitated by "inter-agental concatenations that exceed the previous reach of either party" (Connolly 2011, 27). It should be noted, however, that agency is a process, and as such, it "is never consummate" (Connolly 2011, 27). The way in which activists and advocates join forces and collectively decide the path taken by an assemblage they form shapes the entangled agency as it occurs.

What I have termed *entangled agency* is thus a form of avoiding a unifying message that would be the product of an exclusionary hierarchical decision-making structure. As power structures, networks are less prone to concentrate authority on one or a few actors than on other types of structures, although they nevertheless do the former too. The distributive capacity that they often defend as their defining characteristic is more of a horizon or an ideal to strive for. And yet, the way advocacy networks function is by continuously earning the respect of their members for being a valuable joint effort. This process is closely related to a harmonization with the nonhuman life forms, objects, and other elements involved in the issues that bring all actors together. In the case of the Amazon networks, these elements would include the plants and animals they are trying to protect, the insects they count or study, the droughts or fires that affect forested areas, as well as the roads, townships, and other human-made forms. In such configurations, agency therefore becomes a type of awareness of power.

A few days after the meeting I described above in Alter do Chão, I talked with Tony over a *cafezinho* (espresso with sugar) on the terrace of his house in

Belém. I was interested in knowing what he thought about the clash between the two conflicting visions of IMA members at the launch of a new network. Would he not be worried about what members of other NGOs might think of IMA after having witnessed that open confrontation? Much to my surprise, he told me candidly that he found such tension among stakeholders "actually very productive, because it obliges them to have a dialogue, and in the end I can see that it makes it more effective because they take into account other points of view." As a public disagreement among IMA's staff, however, some of its members feared it would weaken their role as mediators or political actors. Tony disregarded such fears by again considering the positive outcomes of disagreements: "I am in favor of ambiguity; I am in favor of diversity. I don't think anybody is *dono da verdade* [owner of the truth]. Reality is complex, and I think that it is often convenient for people to be in doubt about precisely where IMA stands because there are very different and conflicting interest groups and stakeholders involved, and there is no reason why one needs to burn oneself with one group just to satisfy another. And I think it's honest, we do have differences, they're legitimate and we can handle it."

For Seligman and Weller, humans have always sought to make sense of the complexity or chaos in which we live (Seligman and Weller 2012). A key tool that has been used throughout human history to help us fathom our surroundings is that of ambiguity—be it in religion, law, or social norms. The way in which such ambiguity has been provided with some illusion of order is through repetition (Seligman and Weller 2012, 24), which has allowed rituals to become markers of symbolic significance for collective sense-making. Seligman and Weller argue that institutions such as religion or kingship have historically used their overarching legitimacy as bearers of social order to define the categories that we need in order to deal with life's ambiguities. This was done through performed rituals that were an exercise of notation, which provided meaningful interactions and made sense to everyone. They also suggest that rituals that used to be common are not so present in current public life: "Much of the world is today, for better or worse, much less ritualized than it once was" (Seligman and Weller 2012, 151). They therefore suggest paying attention to "shared experience" as a source of common knowledge that can help societies deal with problems they face. Perhaps advocacy networks represent ideal arenas where such experiences are shared, knowledge is produced, and solutions to problems are found. As Seligman and Weller shrewdly note, "Meaning rests not on the knowledge of 'things' but on the relations between them, that is, between us" (Seligman and Weller 2012, 197).

In the end, the moderate vision prevailed in Alter do Chão, and the network of leaders formed there managed to become influential in Brasília. The meeting I witnessed, as well as subsequent visits to some of those people who were involved, showed me that the process of its formation had started long before my arrival and continued long after I left. The network itself was not static, but rather it adapted to changing circumstances, according to the political and economic context in which it found itself. By successfully weaving a collaborative ethos from its members' input, the conveners managed to enable a functional network. It reached its decisions and executed its actions through negotiations between all those involved. In defending ambiguity, what Tony was stressing was the importance of pursuing a manner of acknowledging the entangled character of the network's endeavor. Indeed, this was a fundamental aspect of its agency.

This chapter considers advocacy networks as having a specific type of entangled agency, which is different from other forms of agency. It is based on the networked disposition of the collaborative work that goes on within advocacy networks. This disposition can vary depending on the interactions that make up each network. I identify two types of networks: harmonious and dissonant. The first are those whose members already share key aspects of their work, which may be ideological, identity-based, agenda-wise, or other. The dissonant are those networks that purposely seek a plurality of groups and organizations regardless of their contrasting viewpoints or agendas, as long as they agree on the issue that brings the network together. Decisions and action may take different forms in each of the two types of networks. But in both cases the capacity of the network to reach agreements over principles and endeavors, which is understood here as agency, cannot be reduced to the capacity of some of its groups or as an aggregation of capacities. The resulting entangled agency is also fundamental for the process of civil becoming, as it in itself constitutes a political experiment of plural negotiation. In this sense, advocacy networks enact a type of democratic decision-making that is based on innovative communication strategies and an entangled agency that seeks to promote a just and fair participation of each of their member groups.

PART THREE

Alternative Performative Politics

5

Marches and Meetings

It is not about those in the north telling those in the south what
to do, but that we both speak of our problems which we may not
understand because we do not share them. If we do not see this
in terms of a dialogue, then I am not interested in being here.

—Ahmed, Algerian activist, member of the FSMed's
International Coordinating Group

As I WALKED through the Mediterranean Social Forum (Fòrum Social Mediterrani, FSMed) event, I felt the wide spaces somewhat empty. The Fira de
Barcelona's buildings are designed to host commercial fairs, where participating companies or groups invest in booths, whose layout may be arranged in
a variety of ways. For the FSMed event, however, these large areas had been
separated into distinct areas for panels, workshops, and conferences. Over
four days, the FSMed event unfolded similarly to other versions of the World
Social Forum (WSF) process, with hundreds of small- to medium-sized
meetings organized by participating groups (called "self-organized events")
and a few plenary sessions (designed by the organizing committee). There
were also group stalls, artistic exhibitions, and political performances as well
as a stage for musical and other cultural events. The purpose was for activists
from around the Mediterranean region to come together in order to establish or strengthen relations that would help their individual struggles. The
small meetings were akin to academic conference panels, as several speakers would take the microphone before the meeting would conclude with an
open discussion for all attendees. The plenaries were more like academic conference keynote panels in that they focused on issues that all the organizers had agreed were the most pressing. They were larger than the previously
described meetings both in number of speakers and audience, and they did
not include time for questions and answers. As with other social forums, the

FSMed also held a final collective march, showcasing the diverse groups that gathered under its umbrella. In all of these activities, *being there* was of the essence. As Judith Butler argues, "forms of assembly already signify prior to, and apart from, any particular demands they make" (Butler 2015, 8).

I consider all of these collective actions as embodied forms of political manifestations. What makes them political is the intent of the FSMed event to debate or express opinions about common affairs with the hope to alter them. Activist groups or NGOs may hold similar events with similar intents on their own, but such assemblies carry a different weight when done in the context of an advocacy network. This chapter is an examination of such embodied political manifestations as collective performances of the political. I include other gatherings that were fundamental to the FSMed process leading up to its four-day event. These were dozens of small- and medium-sized meetings during the organizing process as well as a few large assemblies, public talks, and dialogues, which took place in 2004 and 2005. Although these other gatherings did not have the making of public political statements as their aim, they operated in an environment characterized by a shared understanding of a common political perspective regarding the global status quo and its repercussions in societies and ecosystems around the Mediterranean. All preparatory meetings that were part of the FSMed organizing process, therefore, shared a mission to promote alternative political engagements that would challenge those led by government institutions and economic forces. For my analysis, I categorize these gatherings into three main types of congregation: *open deliberations*, which consist of public meetings; *closed deliberations*, which are private meetings; and *manifestations*, which include public marches and protests. A fourth type not present in the FSMed would be a group's *undertakings*, consisting of their quotidian activities in following their own agenda (e.g., an NGO's educational projects, or a squatters' group squatting). Undertakings can also be collaborative, with several groups participating in the pursuit of a shared goal.

Each of these types of alternative performative politics is laden with symbolism about belonging, aspirations, and denunciations. These performances are thus essential for activists' sense of purpose. In studies about political mobilizations, manifestations usually attract more attention as key practices that illustrate what activists do (Graeber 2009; Juris 2008a). Recent anthropological interest in meetings, however, has extended to those of activists (Corsín Jiménez and Estalella 2017) and NGOs (Rutherford 2004). This points toward a more rounded appraisal of what actually takes place among dissenters. Before expanding on each of the types of congregation I consider relevant for my analysis (open deliberations, closed deliberations, and manifestations), a short exploration of the analytic value of the performativity of assembly in its embodied and time-related dimensions follows.

Judith Butler recently theorized on the performativity of assembly in light of the major mobilizations around the world known as the Arab Spring, the Occupy movement, and the antiprecarity demonstrations (Butler 2015). She argues that the ideals of democracy rely on a "people" in its physical aggregation, which informs the types of freedoms that are held among its core principles—expression and assembly: "If we consider why freedom of assembly is separate from freedom of expression, it is precisely because the power that people have to gather together is itself an important political prerogative, quite distinct from the right to say whatever they have to say once people have gathered. The gathering signifies in excess of what is said, and that mode of signification is a concerted bodily enactment, a plural form of performativity" (Butler 2015, 8). Collective forms of action, therefore, "can be an embodied form of calling into question the inchoate and powerful dimensions of reigning notions of the political" (Butler 2015, 9). Whichever institutional design exists to channel and constrain state power, assemblies can tap into the political simply by coming together.

In theorizing such co-presence, Butler was thinking of marches and protests. But, would this also apply to meetings? I believe so, and I will try to prove it in this chapter. The groups portrayed in this volume are made up of individuals who intentionally seek to play a part in the "dramatization of the political" (Abélès 1988, 391), albeit not as part of institutionalized forms of power. There is a long tradition in anthropology of studies of meetings as significant moments in collective affairs (Schwartzman 1987, 1989; Brenneis 1994; Brown, Reed, and Yarrow 2017). For activists and advocates, meetings are opportunities to establish social links that will allow for collaboration to occur. Meetings are not only about sharing information or defining activities, they are also a way to build a sense of community among all participants. Being there is also crucial to signal commitment, disposition, and engagement. All of these issues have repercussions for the political agendas at hand.

Regarding all types of alternative performative politics already mentioned, it is worth reflecting on their significance for the process of civil becomings. The groups taking part in advocacy networks seek to take advantage of these collaborative efforts not only for the stated purpose that bring them together but also crucially in order to build their legitimacy as members of an active civil society. In part, this phenomenon is enmeshed in an effort to bring about changes in wider social assemblages. In her recent volume *Figurations of the Future*, Stine Krøijer explores radical leftist activists' focus on enacting the future they desire. Activists' performativity, Krøijer argues, is an indication that the body is "the site of politics" (Krøijer 2015, 15). The title of her book stems from her response to Maeckelbergh's interpretation of activists as enacting *prefiguration*, which Maeckelbergh defines as "a practice through which movement actors create a conflation of their ends with their means" (Maeckelbergh 2009, 67). As Maeckelbergh observes, "In my experience as an activist, practising prefiguration has meant always trying to make the processes we use to achieve our immediate goals an embodiment of our ultimate goals, so that there is no distinction between how we fight and what we fight for, at least not where the ultimate goal of a radically different society is concerned" (Maeckelbergh 2009, 66). By this, therefore, Maeckelbergh meant that in acting in a way that is coherent with their objective, activists are bringing the desired future into the present.

Krøijer argues that Maeckelbergh's proposal collapses the temporal distinction between present and future, thus following "an ontology of linear time, which does not permit us to adequately understand the radically open and indeterminate elements of activist practices" (Krøijer 2015, 27). In order to avoid this herself, Krøijer chooses a perspectivist model of time, where "the future is not thought of as a point ahead in linear time, but as a coexisting bodily perspective" (Krøijer 2015, 28). For this, she uses the concept of political cosmology as informed by Eduardo Viveiros de Castro's theorization of Amerindian perspectivism (Castro 1992, 1998, 2004). Krøijer prefers cosmology and avoids ideology, because the left radical activists she followed "generally refrain from providing images of the ideal or utopian society towards which they are striving, which challenges our implicit understanding of politics as a goal-oriented practice" (Krøijer 2015, 39). Krøijer's resulting analysis is a sophisticated application of a theory originally developed as an interpretation of Amazonian people's understandings of time and being.

These reflections provide valuable insights into how activists enact the changes they desire. However, I do not consider them applicable to both sets of networks that are analyzed in this volume. A key element of Krøijer's analysis was the fact that all activists identified themselves as left radical, and that provided a common basis for their reflections and actions. In contrast, the two cases that I analyze in this book—in the Amazon and in the Mediterranean—lack ideological homogeneity. At the most, one set of networks is harmonious insofar as all those participating fall somewhere on what can be considered the left end of the political spectrum.[1] Furthermore, some of the NGOs involved—especially those working on development-related issues—work within an analytic frame of linear time, assuming that the changes they promote are gradual. These same groups also often set up mid- and long-term goals and must work within established forms of accountability to their funders. For these reasons, I do not believe that advocacy networks function with what Krøijer calls a political cosmology.

1. In chapter 4, I differentiate advocacy networks as "dissonant" and "harmonious" according to their internal cohesion. The former strategically include contrasting positions or strategies, and the latter favor a shared form of identification, be it ideological or otherwise.

Rather than seeking a sense of the collective within the networks, what I seek in my study is an analysis and understanding of what takes place in such diverse coalitions. In stressing a sense of civil becomings, what is emphasized in these pages is the ongoing search for legitimation in which these groups engage. The networks I followed served thus as a peer-review system in which the groups involved vouched for one another as a valuable member of what they consider to be "organized civil society." To achieve this, the various performances of dissent I outlined above, and expand on below, are essential. The process just described was particularly interesting in Barcelona, as the organizing mechanism I witnessed was considered by all participating activists as an umbrella organization whose aim was to achieve a single four-day event.

Deliberations

In the eighteen months leading up to the FSMed, the Technical Secretariat met weekly to discuss preparations for the event (figure 5.1). These gatherings were closed deliberations, where decisions would be made regarding practical issues. These meetings mostly consisted of negotiations among participating groups regarding what was urgently needed and how to distribute tasks. I realized, like Helen Schwartzman, that meetings were not so much a source of knowledge about something else (each task or decision), but rather central to the collective organization or social assemblage at play. Schwartzman "came to see that it was the meeting itself that should be the subject of study" (Schwartzman 1987, 272). Numerous sessions consisted mainly of what groups were failing to achieve. An early example for me was in a meeting held in early May 2004, in which Javier argued that it was necessary to strengthen the commissions that had been set up to prepare for specific tasks[2] that would ensure the smooth running of the event taking place the following year: "We seek to evaluate what has been done so far to define priorities, and from these evaluations, committees can get to work in their specific projects, in order for us to finally get to work [*para ya ponernos a trabajar*]." This last phrase was a somewhat passive aggressive indication that, at least for him, there was no evidence of work being carried out by those who had committed themselves to doing it. María then summarized the preparations so far (for example, that the dates for meetings of the International Coordinating Group had been set), and ended saying that the FSMed event would take place in a year and two months, adding, "so we have a rough guide of what we want to achieve in that time." Javier again demonstrated his preoccupation with the committee's apathy in his next intervention:

2. There were five committees, each one in charge of one of the following: the program, logistics and finances, promotion, institutional relations (especially with governments), and media (for example, dealing with journalists and preparing a press room). (See figures 3.1 and 3.2.)

Figure 5.1. Panel at the Mediterranean Social Forum (FSMed) event, 2005. (Raúl Acosta)

> Javier: Committees are not working with the vitality with which they should be. We are expecting fifteen thousand participants. It is vital to strengthen the Promotion Committee, not only to reach organizations in other countries, because we have so far only had local groups from Catalonia join. But there are also two committees that have not been working as was expected, or at least with a diminished participation. It is necessary to debate and reflect to understand why the FSMed is necessary . . . it is not just meeting to meet. It seems that the same dynamic as that of the last four years is returning, that make it difficult.

Javier's accusation was not met with unanimous support. María's immediate response sought to try to contextualize the difficulties some participants were having:

> María: If we compare the current situation with that of last October, we can see that we have been advancing. After [the bombings in] Casablanca, the regime has been repressing all opposition groups. It has not been easy. We need to understand the climate in which NGOs are working, and clarify what mechanisms we propose for working together.

This type of discussion would often take up large portions of meetings, as we grappled with events taking place around the Mediterranean that would affect the groups that we deemed to be potential participants at the FSMed event. For some activists in Barcelona, it was incomprehensible why activists in the southern shores of the Mediterranean had not achieved more. But they appeared to assume that activist meetings could take place without problems, and this was not the case. For many groups in the Maghreb, for example, politically oriented gatherings were dangerous affairs due to state security efforts to prevent them. Anthropologists studying council systems have argued how such meetings rely on power to legitimize their decision-making (Richards and Kuper 1971), so any similar grouping that in itself challenges such status quo poses a threat to established authorities.

Judith Butler's proposal for a theory of assembly argues that public forms of assembly "can be understood as nascent and provisional versions of popular sovereignty" (Butler 2015, 16). In her view, political performability lies at the core of democratic theory. This view coincides with some of the insights of Keane's recent history of democracy, which stitches together a nonlinear narrative about the many actions and ideas that have contributed to current understandings of democracy as a political form of government (Keane 2009). For any polity, social assemblies are symbolically powerful. It is no wonder that authoritarian governments seek to prohibit both large and small gatherings they consider threatening. Large congregations may pose too blatant a challenge to authoritative or powerful regimes. Small ones may provide individuals with an opportunity to develop ideas and ideals that may prove dangerous to the regime. Examples abound of prohibitions of this sort. In democratic regimes, freedom of assembly or of association is considered a building block of a free society. It constitutes one of the basic human rights that is enshrined as Article 20 in the United Nations' Universal Declaration of Human Rights (UN 1948). Civil society organizations therefore use this right in order to pursue their objectives. Both Brazil and Spain went through dictatorships in the second half of the twentieth century. In both cases, groups of individuals who were committed to seeking a change of regime played a fundamental role in achieving the end of dictatorship and establishing periods in which these nations transitioned into democracy. Several of my interlocutors on the FSMed organizing committee had been actively involved in social mobilizations against the regime led by Franco. Javier was one of them.

Bailey referred to politics as having a public face with normative rules, and a private wisdom with pragmatic rules (Bailey 1969, 5). Meetings of alternative political movements thus challenge established rules of institutional governments or established power holders. Bailey himself described politics as enterprise (Bailey 1969, 35), in which a strengthening of a sense of community comes from "an animosity that is cultivated and *focused on an external enemy*" (Bailey 1998, xii). For activists and advocates, this process usually involves an active distrust of the state, which is directed to government institutions. Although democracy is considered a valuable system of government, some activists often reflected on its fragility, especially in light of their own experiences. Two weeks after the FSMed event took place in 2005, a group of secretariat members held a dinner to celebrate. It was in a restaurant overlooking the city's port. After dinner, most of us went to a bar close by, where a few rounds of spirits and beer were gladly shared. Javier asked a colleague: "Do you remember the 23F? We had guns and rifles under our beds, just in case . . . there was a lot of tension in the air." He was referring to the attempted 1981 Spanish coup d'état during which time a group of military leaders sought to take control of the government after the first few years of a difficult transition to democracy. "We were ready for what may happen . . . we were fearing the worst," said the colleague, with a smile of camaraderie. Franco's death in 1975 had meant the loss of the strong man in charge of the military dictatorship that controlled Spain for almost forty years. The subsequent legalization of the communist party and the rising influence of the socialist party were considered outrageous by some military and conservative leaders who sought to regain control after Franco's death. In the end, the attempted overthrow was a failure, and Spain went on to celebrate free elections. "What we learned that day was to never let our guard down," said Javier.

Technical Secretariat meetings, therefore, were about more than just the organization of the FSMed event. They were about shedding light on political and environmental problems around the Mediterranean. Javier often reminded everyone that it was important to recapitulate the relevance of the FSMed, that "it is not just meeting to meet." This phrase was a veiled criticism of what appeared to be a broken synergy, as he and others would tell me at different points that the ongoing meetings appeared to be ends in themselves, instead of amounting to a progressive realization of assignments. As a labor union leader, Javier was all too aware of how meaningful a commitment is. His work in the political organization of the union often included meetings and negotiations, with specific goals of collective action. Workers join unions to achieve certain protections that they would not have if they remained outside unions. In the FSMed, however, the aims were not as palpable. Javier's discomfort resonates with Riles's view of a modern bureaucratic conception of meetings: "a form of interaction that has outputs" (Riles 2017, 182). Whereas social forum documents and proponents emphasize its focus on process, for FSMed event organizers it was important to achieve a set of goals that would expedite the success of the four-day event. Javier called for outcomes from these meetings that were not forthcoming. Riles points out that for many people with such a frame of mind, "meetings without outputs do not produce 'no meeting'—they produce an illegitimate (i.e., pointless) meeting" (Riles 2017, 186).

It was easy for me to understand Javier, however, especially regarding the expectations of participation. For the FSMed to host fifteen thousand participants, there needed to be more done simply to inform people who might be interested. Later in the aforementioned meeting, in which Javier bared his teeth, Víctor raised a related issue. He had recently returned from Istanbul, where he had attended a preparatory meeting of the European Social Forum (ESF) that would take place in London in October 2004:

> Víctor: First, it was agreed that we would have a strong presence of the FSMed at the European Social Forum in London, perhaps combining a conference, a seminar, and a working group. This would help us start the political debate about what is the Euro-Mediterranean Area and the free trade zone.
>
> Second, the Greek delegation (representing the Greek Social Forum), complained about how close the FSMed was to the ESF, in time and resources for many organizations. I present it here so that we consider it.

Víctor was not alone in his concerns. I had heard from several activists that there was a real anxiety among some organizations about a "forum fatigue," as organizations had limited human and material resources to engage in all the forums that were available. This was especially the case when each of the events required several planning meetings and negotiations of their own. It did not help that some groups of influence in the region, such as the Greeks, questioned the legitimacy of the FSMed by saying it was not adding much to what already existed in the ESF. After a diversion into other topics, the conversation returned to the question of the FSMed and ESF:

> Gala: I am worried about the Greek problem [by which she meant the critique about "forum fatigue"].

> Víctor: There will be another meeting in Paris in May to continue preparations for the ESF. Then there will be another meeting in Berlin where the program will be approved. We have asked for a slot in the Berlin agenda for Catalonia, to propose our activities for London, but also to address this situation [the "Greek problem"].

> Javier: These fears are reiterated, and serious. We are also discussing the FSMed's regularity. There's a lot of division in Greece, and in Italy, regarding these issues. I am worried. Those of us who have participated [in the ESF], have noticed that something is going on. As Mediterranean as it could be, however, an ESF is European. About Víctor's

proposal, I think it is necessary to discuss it, but not indispensable. The effort it entails may not be worth it.

Víctor: There is a lack of critical reflection about the Europartenariat process [the Euro-Mediterranean Partnership promoted by the European Union], and about the free-trade area in the Mediterranean. The proposal of both seminaries for London points in that direction, not only to announce the FSMed, but because we want to stir debate. . . .

Darío: Enough with paranoia . . . let those who are interested come. There are already enough materials [texts] about the free-trade area.

Javier: It is good that issues relating to the Mediterranean are to be discussed in London, but I suggest that individual organizations do this by themselves.

The conversation thus referred to a preparatory meeting of the ESF where Víctor achieved an agreement with organizers that the FSMed would be present at the ESF event, while acknowledging the skepticism of some activist groups—especially the Greek ones—about the need for yet another forum. The entanglement of meetings was a bit overwhelming. Each social forum requires a series of preparatory meetings, and, as I already pointed out, it sometimes seems that the aim of one meeting is merely to facilitate another meeting. Some of the most seasoned activists, like Javier, seemed to take the proliferation of meetings in stride—albeit demanding concrete results—and recognized the strategic significance of spreading the word about the FSMed in different encounters. For others, like Darío, who was in charge of the FSMed's web page and finances, there seemed to be too much focus on keeping up certain appearances. Darío was in his fifties and worked in a software company as a programmer. He was in a comfortable socioeconomic position, but often paraded openly his left-wing ideals. He also complained about a lack of visible results from meetings.

Other closed meetings that were crucial for the FSMed organization process were the ones with government officials. These were carried out by one of the committees, made up of four activists and advocates, who were actually two representatives of each of the two camps (radicals and reformists). In some of the gossip sessions after meetings over a beer, I was told that places in such a committee had been hard to negotiate, as they were considered to be extremely valuable for the organizations represented. The committee would meet local, regional, and federal politicians to negotiate the amount and character of support for the FSMed event. This included the venue sites, permits for the marches, issues regarding visa applications by activists from outside the European Union, and other issues. "Whoever meets these high-ranking politicians earns a place in their minds as a representative of civil society," Montse told me, with a wink and a smile. Montse was a twenty-nine-year-old union worker, who was a close collaborator of Javier both in the union and in the Technical Secretariat. In her view, such positioning "may mean easier access for other things later on." These were some of the practices that radical activists often accused reformists of. Although government support for civil society organizations often had official bureaucratic procedures (in the form of grants and other aspects), personal contacts with relevant politicians helped.

One of the meetings I attended was the first of the FSMed's International Coordinating Group, which took place in Barcelona on Saturday, July 10, 2004. The group had recently been formed at an assembly in Cyprus, with the alleged aim of strengthening the planning of the FSMed event across national borders. Participants from abroad arrived on Friday, and they would leave on Sunday. Thus, the facilitators had only one full day to address a long agenda. This was a challenge, especially considering that several attendees had never previously met. On account of this, they had to quickly navigate the uncertainties that are common in initial encounters while covering a number of important issues. The first agenda item was reports from each of the delegations regarding the various committees that had been set up by the International Assembly. The first report was dedicated to the Promotion Committee, which had been tasked with seeking to involve activist groups or NGOs in the FSMed—either by helping its organization process or participating in the four-day event. The reports were to be verbal, and they would be registered in the minutes of the meeting. What took place, however, was a series of discussions that highlighted some of the contrasts between the circumstances and attitudes of the northern and southern groups who were involved in the FSMed process. While some groups from the Technical Secretariat tried to push for concrete agreements and proposals, other participants prioritized establishing trust and understanding. There seemed to be frustration among many participants who felt their concerns were not being addressed:

Ahmed (from Algeria): I speak on behalf of the Berbers.

The problems of the south are different from those of the north. When we speak about promotion, well, it is a work on awareness of civil society through organizations. But it is difficult with many organizations, because many people do not feel these represent them because there is a lot of corruption.

People do not feel they can trust, because whoever will be able to pay for their own [airline] tickets will not be representative. Our meetings should be in the south, because this would initiate a process through which those in the north would listen to our problems in their context and would understand who is being sincere. Organizations that are more dedicated to [political] mobilizations are very busy and cannot really dedicate time to promoting the FSMed.

Alicia (Fons Català): We will talk about the meetings in the south at the end of this meeting, so that there are proposals. It is important that more than saying there should be meetings there, concrete proposals are put on the table.

Asha (Israeli Palestinian): I don't think what is important is that northerners go to the south to do the promotion, but that the southerners do it in our own countries, so that the small groups grow.

Alicia: Southern organizations must evaluate which ones are the important topics for their regions, and invite who they consider relevant.

Salman (Egypt): I am very happy to be here, among friends. I have a lot of experience listening to politicians. When I said at the beginning of the meeting that we should talk about the situation in the south, I did not want to blow it out of proportion. It is not an object of study, but a subject of dialogue. There have already been meetings in Cairo and other places to talk about the Porto Alegre process [the WSF]. But there are problems because civil society organizations that have links to social movements are not grassroots, and it is like Ahmed described in Algeria (it is not worse, but more complex). In our countries, the influence of Europe and the United States of America in [organized] civil society is financial and ideological. This means that organizations constitute an elite and are not linked with social movements. It is difficult to find a way out of this process. In Egypt, we have not been able to break it. I think it is the case in the majority of our countries [in the Maghreb], where any truly dissenting meeting is heavily repressed. We have tried to get out of this situation, but it has proven difficult. We are only starting. I do not want to be a pessimist, because I am not, but this is the situation, and we must take it into consideration.

Ahmed: In order to carry on with other issues, we can talk about this one after lunch, because it is important for us.

Fabrizio (Italy): In Italy we have had two meetings, but we are really delaying the issue [of the FSMed] because we have had many problems that have emerged. For example, we invested in long meetings about the European elections. It appears to me that it is very difficult to achieve an agreement to hold a national meeting towards the FSMed [event]. We may be able to hold it in September. From debates in two meetings, there have been some commitments among a few networks, but we need to evaluate their results.

Alicia: We are now informing, we will debate later. Víctor, do you want to talk about Turkey?

Víctor: Once we finished the assembly of the European Social Forum, there were around 50 people from Turkish organizations left, who were very happy because there was such a good representation from many corners of Turkey. They were truly happy.

Fabrizio: The proof of success was the party afterwards. . . . [general laughter]

Giovanna (Sicily): In Sicily there was a first meeting towards the FSMed. It was held by a network of unions and churches on the forum process. Grassroots. One was the non-Catholic church that has worked against the mafia and other issues. We have convened a meeting for September in order to assemble all the social movements of the island.

Asha: After the meeting in Cyprus, we went back and thought about working on two levels: within Israel, and outside. Twenty-two organizations took part in the first meeting. It was less than we expected. Some of those who did not attend, could not do it because they are very small and without resources, and also very difficult to reach. We are also trying to have an Arab Social Forum, and an Israeli one as well, so they invited me to also talk about these. I have also been invited by women's [rights] groups.

The new group was created with the idea of opening up decision-making to all the participating regions. The different expectations participants had of its first meeting revealed the rifts among regions and localities. While members of the Technical Secretariat assumed the gathering could be a swift revision of outstanding tasks in order to ensure their completion, many others sought first to establish some sort of rapport, and thus trust, among the groups. Other groups were somewhat suspicious of the Catalans' insistence on an "efficient" meeting. The rounds of discussions, therefore, included many comments that were more about how certain groups perceived the whole FSMed from a distance. At one point, for example, there was a complaint by someone from the Technical Secretariat that a bulletin that was originally proposed as a communication tool for the FSMed process had not been completed because no groups would send articles or texts. To this comment, Ahmed reacted angrily:

Ahmed: It is not about those in the north telling those in the south what to do, but that we both speak of our problems which we may not understand because we do not share them. If we do not see this in terms of a dialogue, then I am not interested in being here.

The point is to say that we are not alone, that we accompany each other, and that solidarity and fraternity actually exist. The FSMed must look for its identity, what it is really about. I need to know how to explain it to the people at the base. Their problems are very clear: corruption, poverty . . . issues like the International Monetary Fund, etc., do not interest them because they are very far [from people's everyday lives] . . . but we need to understand their worries.

Ahmed thus referred to the sense of representation that was not often discussed but was always present in the FSMed preparatory meetings. If the first meeting of the International Coordinating Group's purpose had been to serve as a place of convergence for the various diverse struggles of the region, just how did it strike a balance between analyses achieved through expertise and quotidian struggles of populations dealing with injustice, poverty, and other ailments? This discussion I quoted at length above hammered home the thrust of participants' ideas of their role as organized civil society. Over dinner, Salman, who was widely seen as a respected elder in the international committee, aired his sense of unease about the process, to Javier and others around him: "These meetings are too orderly, too clean. They are artificial. We are not able to feel at ease, to get to know other participants . . . time is restricted, always, but it doesn't feel that we are included in deciding how these meetings are run." He was being open and honest with Javier and other activists from the Technical Secretariat as well as with individuals from other countries. His comment pinpointed a tension I had noticed throughout the entire organizing process. The Catalan committee sought to achieve an "efficient" organization of the Forum, as several comments revealed. The purpose of the social forum process was precisely to establish common ground among a wide array of struggles in order to enable new forms of collaboration, mutual understandings, and help groups push for progressive social issues to be included in political agendas. This required participants to enact their roles as both focused on their specific group's program and committed to collaborating with like-minded groups.

The need for relational forms of legitimacy was at the root of the use of assemblies as key nodes for decision-making. As I explained in chapter 3, however, the actual organizational arrangement of the FSMed process diverged somewhat from its stated aims. This meant that the assemblies, a type of open deliberation, were not actually as important as they were made out to be. Assemblies worked as town-hall meetings with dozens of participants who could pose questions and reach decisions about the process. Among the FSMed organizers, assemblies were conceived as spaces for collective deliberations on organizational or constituent issues. The most common organizational issues included crucial dates, distribution of events (for example, plenaries), and principles for how to approach governments for support, among others. Constituent issues were the ideological or practical propositions shared by those taking part, which could be stated as being central to the FSMed ethos. Groupings such as the Technical Secretariat or the International Coordinating Group, that is, those that held closed deliberations or private meetings with a few participants (never more than twenty), were tasked with dealing exclusively with the practicalities of the FSMed process, while the important decisions were left to the assemblies, that is, groupings that held open deliberations. This meant, in theory, that the process was as horizontal and democratic as possible. In reality, however, all of the crucial decisions and negotiations actually took place during the closed deliberations of the Technical Secretariat. Thus, a limited number of activist groups and NGOs wielded power over resources and far-reaching decisions. Even the International Coordinating Group had only a symbolic role.

There were two types of assemblies as part of the FSMed preparatory process: a general one and another only for Barcelona. Assemblies were open spaces where any group or individual interested in the FSMed could attend and participate. The general one was open to groups from around the Mediterranean, and the Barcelona one welcomed those based in either the city in particular or Catalonia more generally. Mediterranean assemblies were held in Rabat (Morocco), Málaga (Spain), Marseille (France), Istanbul (Turkey), Cyprus, and in Barcelona. I attended one Mediterranean and one local, both in Barcelona. Going through my notes of these meetings, I had a sense of déjà vu regarding other meetings, including those of the Technical Secretariat: I kept encountering complaints about committees not doing enough; pleas for more promotion in order to ensure more participation; and litanies about upcoming meetings in various places (either from different forums, such as the European Social Forum, or from specific networks) where the FSMed could be further promoted. In both assemblies I attended, the logic appeared to be that members of the Technical Secretariat informed assembly attendees about how preparations were going so far and took votes on issues that had already been negotiated and agreed among members of the Technical Secretariat. When someone sought to propose a different date for the FSMed event because of delays, especially among the French movements who had not been very involved, Javier said, "We cannot keep playing with the dates [changing them]. We should clarify in our minutes." The Technical Secretariat had struggled to find dates where a suitable and affordable venue would be available. Once they had secured the Fira de Barcelona (a large complex dedicated to business fairs, co-owned and jointly managed since 1932 by the municipal government of Barcelona, the regional Catalan administration, and the local business chamber), they needed to stick to it. But this was easier to do within a smaller group, such as in the Technical Secretariat, than it was in a large assembly.

Manifestations

A different type of alternative performative politics is that of public protests in the form of marches, sit-ins, or some type of enactment to make statements of disapproval or objection to something. Studies of activist dissent have often focused on its most visible form: that of manifestations, or public protests and marches. Marches and protests often rely on some ludic elements, such as street theater. These have a specific aim: they serve as scripted and staged presentations (Shepard, Bogad, and Duncombe 2008). Manifestations are often considered by scholars to be the most significant kind of activist event, which comprehensively convey what activism means (Graeber 2009; Juris 2008a). Similar to how government officials use public ceremonies and other performances to continually legitimize their rule, those who challenge that very authority also use public performances. In dictatorships, violence is used to ensure the type of social order deemed appropriate by the regime. In other types of government, the threat of violence is often used with similar aims. Foucault referred to "governmentality" as an internalization of the rules through which a population constrains its conduct (Foucault 2008; Dean 2013). If people assume the status quo as an unquestionable arrangement, then there will be not only a lack of contestation but a lack of desire to contest it as well. The fact that groups actively oppose a certain arrangement emboldens them to publicly manifest their disagreements.

Each of the participating groups in the FSMed Technical Secretariat had a particular history of political engagement that marked their attitude toward public manifestations. On the one hand, reformers were usually weary of appearing angry or incensed and would opt more often for a softer stance. This was due in part to their disposition to sit down and negotiate with government officials to achieve the changes they rooted for. Radicals, on the other hand, were adamant about showing their suspicion of state officials. In their eyes, several government institutions—such as the police, for example—could easily be used against "the people." Some, such as Javier, vividly recalled experiences in which democracy lived under great threat. Javier's memories of the Franco dictatorship and the attempted coup d'état mentioned above shaped his visceral distrust of the Spanish state. During the dictatorship, Javier had worked in an outlawed labor union called Comisiones Obreras (Confederación Sindical de Comisiones Obreras, CCOO). After the fall of the Franco regime, CCOO was legalized and became very popular. In Javier's opinion, it grew too big, which meant that it "ended up working against the interests of the workers." In 1985, he was part of an effort to launch an alternative union "with a less compromised defense of workers' rights" called CATAC (Candidatura Autònoma de Treballadors i Treballadores de l'Administració de Catalunya). In a conversation we held in his office, he told me about this different union: "We did not want to be a corporatist union, but one of class, with clear political proposals and positions regarding workers' problems." It was dedicated to government workers in the Catalan administration. In 1997, he helped form a network of independent unions called IAC (Intersindical Alternativa de Catalunya), with the purpose of joining efforts with social movements and other groups that opposed neoliberalism. This profile made Javier the visible head of the "radical wing" among the groups participating in the FSMed organizing structure. Participants in this alliance mostly came from trade unions and social movements.

The other informal but influential coalition within the FSMed organization structure was what I identified as the "reformer" or "moderate wing," which mostly consisted of NGOs. Its visible leader was María, who, as I have mentioned, was also part of an NGO dedicated to defending minority languages (although most of its work was devoted to Catalan). While the radicals were fond of public marches and protests, the moderates preferred to negotiate with public officials in private meetings. All groups, however, engaged in all types of alternative performative politics at one point or another.

Perhaps the most significant protest I witnessed in the context of the FSMed was the closing march of the event itself (figure 5.2). The order of participating groups had been previously determined as part of a performance we had carefully choreographed because we wanted to avoid friction between acrimonious groups (for example, the Saharawi delegation was placed far away from the Moroccan one, as reported in chapter 3) and hurt susceptibilities ("But they have contributed so much time to the FSMed! They deserve to be up front!"). Once everyone was ready, Javier gave the signal for the march to start. Several groups of drummers enlivened the atmosphere with festive tunes. Some people around them danced. Many flags and smaller banners were waved. The World March for Women had their own colorful props of large swaths of cloth that its participants moved to simulate waves. The closing march was assumed by all participants to be an integral part of the FSMed's essence, following in the tradition of the WSF. All activist groups involved took part with objects, signs, or colors relating to their own agenda. The objective of these marches is for groups to gain visibility—to make the public more aware of the wide variety of demands, proposals, and criticisms that the participating groups hold. The banners and props activists held often had clear messages about their campaigns—though sometimes the messages were cryptic. In either case, they presented photo opportunities that would improve the activists' visibility.

Figure 5.2. Mediterranean Social Forum (FSMed) banner at the march to mark the end of the FSMed event. The order of participating groups had been carefully choreographed to avoid frictions between them and ensure a smooth march. (Raúl Acosta)

The march lasted for almost two hours, following a path agreed on beforehand with local police. The demonstration was quite festive. Its carnivalesque atmosphere made it akin to a number of other events, including: the "reclaim the street" party in Denmark that Krøijer documents (Krøijer 2015, 184); the various sites of anticorporate globalization protests that Juris details (Juris 2008a); and the many direct actions Graeber documents (Graeber 2009). The cheerful atmosphere stood in stark contrast to the range of difficult topics many of the participants were confronting, like denouncing torture of political prisoners or the lack of drinking water for thousands of communities. Such disjuncture is often the case in marches of this nature. Activists require this sort of therapeutic, trance-like moment in order to cope with the negativity of their daily dealings. In dancing and having fun, participants are not denying the heavy hearts they often have. They are celebrating the fact that together they are trying to make a difference. In the case of the FSMed march, as the crowd advanced along the avenues, the festive atmosphere crescendoed. The march ended back at the Fira de Catalunya, with Javier and Asha speaking on the stage. Javier provided a summary of the FSMed as a story of success, especially referring to its numbers—of meetings, participants, languages translated—and of its significance for the region's progressive movements. Asha said it was a step in the right direction, and that there should be more links among all the participating groups and those who could not be there. Following their speeches, musicians played for the rest of the evening, during the Forum's closing party.

The performative value of marches and protests is directly related to their symbolic significance as social assemblies. For example, government officials rely on carefully designed ritualistic parades and events to highlight national themes and display national symbols. The gathered crowds demonstrate the legitimacy of their rule. As a symbol, crowds are akin to massive armies: they have come out in throngs to support their "king." The number of supporters symbolically equals the extent of the leader's military prowess. The same symbolism is at work in protest movements. The larger the crowd willing to stand up to a regime or a powerful figure, the stronger the challenge is. If bodies already speak individually (Felman 2003), they also speak in aggregation (Butler 2015). Protest movements allow for the visibility of collective aspirations for change in the societies of which they are a part.

Krøijer aptly calls marches "intensified moments of political action and protest" (Krøijer 2015, 31). Together with other forms of public protests, they are occurrences of "social effervescence" that Durkheim identifies in moments of political upheaval and revolution and as a basis for the "religious idea" (Durkheim 2001, 158, 64). For Durkheim, such intensified moments of collective life are significant because "vital energies become overstimulated, passions more powerful, sensations stronger" (Durkheim 2001, 317). Because they are so intense, these moments are sites of social change. Durkheim was interested in how the religious idea emerged and took shape. His theorization on it has helped others make sense of the emergence of the political as a form of secular religion. Building on Durkheim's work, Victor Turner paid attention to religious rituals, which he considered as processual and which he analyzed as performances or dramas (Olaveson 2001, 92). These considerations are useful for investigating activists' cyclical use of protests and marches as well as their use of meetings and assemblies. Instead of classifying them as ritual, however, I prefer to emphasize the symbolic character with which they embed their actions. In this, I follow Wendy James's preference for avoiding the term *ritual* and using *ceremonial* to refer to the highly symbolic relations that characterize human bonds (James 2003).

Turner elaborated on Gennep's (Gennep 1960) processual view of rites of passage, turning them into analytic tools that he found useful for examining collective performances. Turner divided the ritual process into three phases: separation, transition—which Krøijer describes as "a state of social limbo out of secular time that generates a strong sense of communitas among participants" (Krøijer 2015, 28)—and reincorporation. Turner emphasized the liminal as transformative. Krøijer argues that for her purposes, "the ontology of linear time underlying the ritual process, even in Turner's version, is a major stumbling block" (Krøijer 2015, 29). I agree with her insistence on nonlinearity in studying protest, as I believe that instead of following a path toward the production of something, activists' performances represent an ongoing state of emergence. It is in the various forms of manifesting their dissent—either through collaboration with state governments, through outright criticism of official policies or projects, or through proposals for alternative ways of dealing with problems—that activist groups embody the sense of what they understand as an actively organized civil society. I do not emphasize an interpretation of symbols—either in acts or objects—in my analysis. Instead, I place emphasis on interpreting the act of coming together to show unified forms of dissent in its variations as a self-fulfilling prophecy of civil society.

With the fall of the Soviet Union and the fall of the so-called Iron Curtain, the early 1990s witnessed an increased emphasis on civil society by governments of Europe and North America (Gellner 1994). Struggles within the former Soviet states (e.g., the Solidarity movement in Poland) became models for activists and policy makers around the world of grassroots practices of democratic aspirations (Cirtautas 1997). Reflecting on the concepts of revolution and agency, Piotr Sztompka developed the framework of social becoming (Sztompka 1990) with the purpose of providing a structure that stresses variations in a nonlinear fashion (Sztompka 1991). He argued that such a framework was necessary because much thinking about revolution and agency ended up reproducing a developmentalist framework that reified a linear narrative. In his proposal, "Social becoming represents the incessant dynamics of producing society by society" (Sztompka 1990, 136). His sociological approach resembles Durkheim's reflections on the fact that the sentiments on which the group is based are reaffirmed through the symbols of ritual.

As with other social forums, the FSMed process was open to any activist group that wanted to participate. Especially because of the authoritarian regimes that abounded at that time in northern Africa, the organizing committee dealt with security concerns about infiltration through a systematic mutual evaluation. That is, legitimacy was awarded by groups that were recognized by others as legitimate. This form of peer review was criticized by some in the organizing committee because of a perceived bias in favor of those organizations from the south that had already been working with European partners, thus perpetuating privileges that came with certain benefits (e.g., money for projects, travel). In all of the discussions in the Technical Secretariat where evaluations were regarded, however, the constant operative public presence of certain NGOs or activist groups ensured their consideration for participation. That is, the longer a group performed their own dissenting agenda openly and for others to witness, the easier it was for other groups to testify to others' commitment and work. In the end, the networks of participation relied on decisions of individuals who formed each of the groups. This chapter therefore ends with a reflection on how personal performances of dissent among members of the Technical Secretariat shaped its path and the frictions among its members.

Alternative performative politics come in different formats and arrangements. The typology that I have defined—open and closed deliberations, manifestations, and undertakings—is helpful for disentangling the strategies through which individuals seek to conduct their activism. A focus on performance "provides a frame that invites critical reflection on communicative processes" (Bauman and Briggs 1990, 60). It is through embodied action that individuals collaborate to achieve joint goals. The FSMed is different from other types of activism in that its purpose is not simply opposing something, such as the anticorporate globalization activists Juris followed (2008b). If the sharp dissimilarities among participants in the FSMed are an example of what takes place in other social forums, the particular contrasts of the Mediterranean (especially between Europe and the rest) stress the frictions that characterized the organization process and that prevented further editions of the FSMed event from happening.

Juris has suggested that through protest performances, activists enact their view of the political (Juris 2008b). Negotiations, agreements, and conflicts were part and parcel of the interactions among organizers of the FSMed event. The ceremonial character of quotidian meetings and special occasions was fundamental for the ongoing collective manufacture of a vision of civil society across the Mediterranean. The mutual recognition and legitimation that activist groups practiced was not simply useful for themselves, but crucially was also a guide for the Spanish government, the European Union, and the international community (regarding development agencies, foundations, and such). By acknowledging some groups as trustworthy and others as less so, the classification was carried out in a way that provided a means for the construction of recognized civil society groups across the Mediterranean. This process was not just from north to south, but also north to north, and south to south. This was not so much the case for south to north, as the power imbalance regarding resources and experience (especially as regards the Social Forum framework) was, from the outset, legitimized in northern-based networks even though it was started in the south (Brazil) (People's Global Action, in Geneva, Switzerland).

In her study of the alterglobalization movement, Maeckelbergh focused on democratic practices within the activist networks (Maeckelbergh 2009). She refers to the "evolving political process" (Maeckelbergh 2009, 30) through which the movement goes. I argue here that such a dynamic consists of performances whereby groups seek recognition as dissenters who earn their right to be considered legitimate members of organized civil society. This recognition, although never official or legal, brings with it certain benefits, which can be material (e.g., access to resources of governments and foundations) or collaborative (e.g., being invited to projects or committees). This in turn results in soft-power gains through which groups gain influence. Such production of sway is an ingredient of civil becomings, as groups want power holders to recognize them as having a pivotal role to play in organized civil society. In this chapter, I have not delved in undertakings, as it was not central to the FSMed process. Undertakings can be manifold, as there exist organizations that are dedicated to a wide variety of issues. In the next chapter, I address one type of undertaking that I consider paramount to advocacy networks: knowledge production.

6

Knowledge, Science, and Legitimacy

Scientific legitimacy is very difficult to gain and very easily lost.
—Nathan, senior scientist at IMA

As WE WALKED through a Brazilian forest, Nathan explained what we should be looking for: "There are two theories: one says that the tropical rain forest lives in a system of co-dependency or community; and the other that it is rather coexistence, an accident that happens to be in the same place and survives. There needs to be a model of interactions between trees. What I want is to have information for our biomodel. For example, when one tree falls, how long does it take for it to turn into CO_2?" It was still early on a sunny day, and we were walking through an area that just a few months ago was burned in a controlled fire as part of an experiment by staff of the NGO I name Instituto do Meio Ambiente da Amazônia (IMA). Nathan was one of IMA's senior scientists at the time of fieldwork, and with him were Janet, Tiago, and myself. Janet was an American doctoral student whom Nathan supervised at Yale University, and Tiago was a Brazilian employee at IMA who was keen to start his doctoral degree. For scientists like Nathan, the forest is both their laboratory and their classroom in which they carry out experiments, take measurements, and make observations for further analyses and lessons in scientific inquiry. In Nathan's case, however, scientific research was not his only motivation for being there at that time. As part of IMA, he seeks to play a role in shaping the future of the region through advocacy, with the help of scientific analyses. As it happens, a main preoccupation for environmentalists is fire in the Amazon forest; it not only destroys large forested areas, releasing carbon dioxide into the atmosphere, but it also becomes uncontrollable very easily. Because fire is one of the traditional ways by which locals clear patches to grow their crops, it is easy to blame related devastation on accidents that happen when individuals seek to clear even small forested areas.

In this chapter, I set out to expand on an increasingly common type of advocacy network undertaking: knowledge production. For numerous advocacy networks, it has been particularly crucial to not only achieve good communicative practices (as explored in chapter 3) but also to produce original information and analyses with which to carry them out. In the Brazilian Amazon, this was of particular importance due to the role of accumulated local knowledges and improved understandings of how the region's ecosystems respond to various types of extraction, urbanization, road-building, and other external influences like changing weather patterns. I consider this a type of alternative performative politics because the process of knowledge production is performative. In the case of scientific analyses, as I will show, those involved also perform their tasks in peculiar ways in order to achieve and maintain the necessary legitimacy in order for their analyses to be taken seriously. This meant also engaging in dialogues with indigenous groups as well as other groupings like the seringueiros, to achieve productive dialogues between contrasting forms of knowledge.

Environmental advocacy is an arena where autonomous collectives strongly contest decisions by governments. A wide variety of groups, from small informal activist collectives to large transnational NGOs, often cooperate in networks through which they hope to achieve policy changes. Collaboration networks have become necessary spheres to produce symbolic capital in order to achieve political influence. In such settings, negotiations over various forms of knowledge have become a common currency used to produce legitimacy for the collective efforts. One dictum Nathan shared with me about the difficult balance that IMA sought is that "scientific legitimacy is very difficult to gain and very easily lost," as he told me while driving to Canarana, in Mato Grosso, on a bumpy road of muddy red clay, surrounded by tall trees on both sides. If they lost the legitimacy they have worked so hard to earn, their sway over government officials would also diminish. In successfully balancing scientifically produced knowledge with that of the lived experience of Amazon inhabitants, environmental advocates have sought to establish a form of "technomoral politics" (Bornstein and Sharma 2016). Bornstein and Sharma define this concept as a mixture of technocratic languages of law and policy with moral pronouncements. In this chapter, I argue that the effective use of knowledge (both scientifically produced knowledge and that gained by the lived experience of local inhabitants) is used to earn the necessary legitimacy for achieving influence with governments and international agencies. It is also a key ingredient to hold together some of the networks through which activists and advocates work. This type of undertaking is therefore crucial for civil becomings as it provides networks with original perspectives to use in their campaigns, which already combine vernacular

Figure 6.1. Nathan, senior scientist at the Instituto do Meio Ambiente da Amazônia (IMA), and Janet, a doctoral student, take photographs and measurements for a long-term study of the forest in the Fazenda Tanguro, property of the Grupo Maggi, in the state of Mato Grosso. (Raúl Acosta)

values and cosmopolitan principles. This in turn increases the legitimacy of the networks in the eyes of other actors.

The experiment our small, aforementioned group was witnessing in walking through that patch of forest was long-term (figure 6.1). Those involved in the experiment were comparing three areas that had been burned periodically by the research team, one more often than the other two. I was shown photos of how this happened, with many people taking care to keep the fire under control, and firefighters ready to act in case the flames started spreading. Although the experiment was well underway, it was still a work in progress, as some related decisions had been made as recently as the day on which we took our walk. Five hundred traps were placed throughout the area to study the seeds that would fall on their nets. Every two weeks, their contents were collected, counted, and then sent to the lab to measure their freshness, alcohol level, or dryness. This would help researchers understand the effects of fire on the Amazon biome. A team of technical staff was in charge of all this systematic work, while the scientific staff visited the site to supervise the work, make adjustments, and revise the data collected. "This is about how the forest is processing energy," Nathan explained. The need to understand the effects of fire are partly due to a preoccupation with resilience, or how easily an area exposed to fire can bounce back. During my time in Brazil, I read a report in a newspaper indicating that the largest source of air pollution in the country was the CO_2 resulting from fires in the Amazon to expand livestock and farming.[1] Fire is thus a major concern in the Amazon. Numerous NGOs are carrying out different projects to try to address the issue. IMA takes part in a network of NGOs and social movements that seeks to coordinate efforts to make their campaigns more effective. Its scientific contribution to such efforts is valued by other participant organizations, funding agencies, and government offices.

IMA's logo on its web page, its vehicles, its publications, and in other media is an image of four large green leaves or trees. Each one contains within it a white silhouette of either a fish, a person, a mammal, or a bird. It stands for the balance sought between all life forms within the Amazon forest, which is the basis of socioenvironmentalism. This logo is placed at the entrance of the Brazilian NGO's headquarters in Belém, the second largest city of the Brazilian Amazon, as well as at other offices and research units in Santarém, Brasília, and other places. Not many NGOs working in the Amazon are based in the region they study. Instead, their headquarters are usually based in the largest cities in southern Brazil (Brasília, São Paulo, and Rio de Janeiro). This is one of IMA's distinctive features, which has helped its standing among regional social movements.

Once we returned to the experiment site in Mato Grosso, Nathan, Janet, Tiago, and I read signs of the processes of the forest as we walked along. Tiago and Janet wrote down what they saw, often with pointers from Nathan. "Charred trees with green leaves are signs of survival," Nathan told me. Other trees were harder to read: were they alive or dead? Shoots that filled cleared areas competing for sunlight were a measure of potential recuperation. All this was to be counted in a periodic census so that IMA could have a clearer idea of the development of the area. It was only one of the sites where IMA carried out experiments. These provided information for IMA's scientists to produce publications and to substantiate the organization's advocacy.

As a scientific NGO, its work is mostly dedicated to carrying out research projects and using that information for advocacy efforts. The legitimacy derived from publications in top peer-reviewed journals and international recognition of its staff's scientific analyses is crucial for successful negotiations with foundations, aid agencies, and government officials. A large proportion

1. The second source was methane from the country's livestock. Much deforestation occurs to clear land for pasture, as farmers seek to expand livestock rearing (Hecht 1993).

of IMA's scientific staff is affiliated with various universities, where they supervise graduate students or teach courses. Its four founding members—who at the time of my fieldwork were in key positions within the NGO—all have doctoral degrees and are affiliated with universities. Pedro, a Brazilian, received his doctorate in ecology from the Universidade Estadual de Campinas, in São Paulo, in 1998, and taught and supervised doctoral students at the Federal University of the State of Pará (Universidade Federal do Pará, UFPA), in Belém. Juliano, the other Brazilian, received his PhD from the UFPA in 2003, where he is now a full-time lecturer in law. Nathan, an American, is affiliated with Yale University, which awarded him his doctoral degree in forest ecology in 1989 and where he supervises Janet's doctoral research. Tony, an American married to a Brazilian, was awarded his doctorate in geography from the University of Wisconsin–Madison in 1989 and is now affiliated with the UFPA. All of them also supervise students from different universities and participate in scientific committees of various organizations. This combination of expertise allows for a blend of resources and flows of new ideas and people. IMA also welcomes other researchers and helps them with technical teams in the areas where they already have research units.

While we were having a beer after a day spent at several experiment sites, Nathan explained to me what drove him to seek the formation of an organization of this type: "When I was doing my doctoral degree, I realized there was a poor understanding of the Amazon forest. Most of the data used in academic papers about the Amazon actually came from research carried out in Costa Rica or other forests." He teamed up with other researchers who shared his interests and perspective, and they founded IMA. Tony had also been interested in using research for advocacy. Before starting up IMA, he had participated with other researchers in the foundation of a similar NGO, IMAZON. On the terrace at his house in Belém, he told me that from the start he was not convinced that IMAZON's efforts were amounting to something: "I had the feeling that IMAZON followed a model that I don't believe is efficient: provide scientific information to feed policy makers, and the policy makers then make the necessary changes." Tony's approach, by contrast, was to encourage the direct involvement of grassroots organizations in research projects. "That is what IMA has achieved," he said, adding, "Our end is change. Policies are another means to an end, but if those policies don't result in change then all you've done is change the policies."

The search for change, however, requires more than scientifically based proposals. In particular, it demands a lot of political tact. Nathan and Tony knew they had to be particularly sensitive about the topic of foreigners working in the Amazon. The region's complicated history with foreigners explains the area's general and persistent distrust of them. Fears of invasion that encouraged the military dictatorship's colonization drive have given way to apprehension about biopiracy. Nathan laughed it off, and he and others told me during a costume party that a few years back he had dressed up as a "bi-opirate." Other reactions against foreigners have to do more with a strong sense of nationalism rather than a specific fear. Large NGOs who are looking into setting up projects in the area must have a Brazilian chapter of their organization to be taken seriously by Brazilian environmental organizations and policy makers. Because of the distrust of foreigners and the area's nationalistic pride, I was told that the American researchers in IMA try to keep a low profile. They need to highlight team effort in their projects while downplaying their individual roles. Tony had ruffled a few feathers in 2003 when he declared to one of the most widely distributed magazines in Brazil that "the Amazon will be occupied." This explosive phrase sparked an outrage not only among government officials but also among other environmental groups. On the one hand, members of the Brazilian federal parliament summoned him to a private meeting in Brasília to clarify his statements. On the other hand, some environmentalists condemned, in public statements and newspaper articles, what they perceived to be an accusation of their work. At the heart of the controversy lay a phrase that appears to be quite descriptive: "We are in an apparently irreversible process. Environmentalists spent a good part of the nineties speaking about the need to preserve or stop and reverse the deforestation rates, and this did not help at all. We lost time that could have been used to think of a sustainable way of developing the region."

Some of Tony's critics raised his American citizenship in their condemnation of his interview. Renata, a researcher who at the time of my fieldwork was the IMA executive director (a rotating post) remarked on Tony's efforts to clear his name and that of IMA's: "He left a very good impression of IMA's seriousness and professionalism . . . they saw that what he had told [the magazine] was not meant as a support for deforestation, which had been one of the misunderstandings." Tony's cool-headed defense of his statements made visible IMA's pragmatist approach to the Amazon's problems—one that swerved significantly from that of so many other groups. Although the episode with Tony brought this into sharp relief, Nathan had already explained to me the difference between their work and that of other activists. Unafraid of tackling controversial issues, Nathan suggested that what was taking place in the Amazon region was a new agricultural revolution, where large-scale agroindustry could exist side by side with small farms and protected areas such as extractive reserves: "There are natural synergies out there with soy farmers who want flat land with no water, and farmers who want hilly land with streams. And such is the natural division of the landscape. We are just at the beginning of that."

Because this perspective was not shared by all of IMA's staff, let alone by the communities with which IMA worked, it prompted debates that reflected the tensions that defined a contested region in Brazil. Tony told me that "we are part of a major regional experiment in physical geomanagement, and I don't think there is another one like it." He was adamant about a major redistribution of land for the purposes of increased productivity and environmental protection. In his view, such a reorganization would ensure not only the protection of large swaths of rain forest but also increased profitability for all stakeholders. IMA's relations with large agribusinesses such as the Grupo Maggi (the world's largest soy producers) were perhaps one of the most mentioned issues by other groupings who accused IMA of using its role to influence neoliberal policies. After all, the fire experiments described above were carried out on a large farm owned by Grupo Maggi in Mato Grosso. "It is a relief [to carry out IMA's experiments within Grupo Maggi's ranch], because in other areas where we have left equipment, it has been stolen, whereas here they have security and all our equipment is safe," Nathan told me while we were driving the thirty-five kilometers between our accommodation and the farm's central offices to check our email. The office was the only place where there was an internet connection. The relationship from which IMA benefited, however, was the source of frequent heated debates, even within IMA itself.

While the scientific moderate group that comprised the majority of IMA could see the benefits of the relations, a second group—the vociferous left-wing radicals—struggled with such a partnership. These two groups reflect a clear contrast in ideological principles and practical matters central to IMA's work. All staff, however, appreciated the constant dialogue that each project stimulated, which included stakeholders like small farmers or fishing communities. It appeared to me that promoting inclusive dialogue was one of IMA's main missions. Tony told me in an interview, "If you have a polarized debate, the radical elements dominate and the interests of most people are lost. The aim must be to separate the reasonable people from the unreasonable people and have that core expand, so I see our approach has got to be to help facilitate that dialogue and interaction." Nathan said it was all about "leveling the playfield" among the stakeholders. He insisted that IMA's role was more that of a mediator who should encourage productive dialogues.

Both Tony and Nathan hinted at ideological differences within the IMA team, which sometimes rose to the surface in private meetings or even in public ones. During my fieldwork, I was witness to several of these heated debates. For this reason, elsewhere I referred to IMA's work as the "management of dissent" (Acosta 2007), which its members embarked on with stakeholders and within IMA's own ranks. The contrasting positions reflect underlying tensions among Brazilians, whose hierarchical society remains one of the most unequal in the world. Socioeconomic inequalities are also evident within IMA. The house where its headquarters are based is a grandiose, late nineteenth-century French-style house. It is in a wealthy area, and many of the nearby mansions testify to the neighborhood's past splendor. Its owners are descendants of a rubber baron, and one of them, Laura, worked in IMA as an associate researcher while doing her doctoral research at the time of my fieldwork. Laura told me about some of her childhood memories growing up in that house. She was seldom seen in IMA's offices, but nevertheless used the best room, which she shared with two other researchers. Her family's history and networks informed her opinions on the region's problems and potential solutions, one of which pertained to the value she placed on ownership. As an example, one evening, Laura invited me to a friend's party in his penthouse in Belém. As is common throughout Brazil, apartment buildings have a security gate with guards, to stave off robberies. What surprised me on that occasion was to find the penthouse was reminiscent of a Swiss chalet, all made of wood with tilted ceilings. It seemed out of place in a city at the edge of a forest. The owner of the penthouse had only recently come back from a long trip to New York. His family had been large-scale cattle ranchers (fazendeiros) for generations. Laura told me on several occasions that before trying to stop people from clearing away the forest to raise cattle there should be an effort to understand why being a large landowner was so well regarded. "It's a matter of status," she insisted, "because even if it's no longer profitable and if they could invest in other businesses, it provides them with a lot of prestige to say 'I am a fazendeiro.'"

The Brazilian admiration for cattle ranchers was made evident to me in an unexpected way. On one of my trips deep inland into Pará's forest, after driving for a few hours along a dirt road, I stayed overnight in a small house. It was where Osvaldo, one of IMA's technicians, lived with his wife and young son. Osvaldo assisted Mariana in her work as political operator in the region. He frequently traveled with her to visit small farmers' unions and other local associations to discuss environmental problems brought about by cattle ranching. It was thus important that he live in the area in order to know all the stakeholders they liaised with and to visit them regularly. The area, as with many others in the Amazon, was only connected to larger urban centers by dirt roads. In part because the area was so remote, it was subject to a considerable amount of illegal logging and cattle ranching. The negotiations he undertook with Mariana were aimed at improving the regulation of ranching in order to stop ranchers' disordered—and illegal—expansion. It thus came as a surprise to me when I overheard him playing with his two-year-old son while I was driving the pickup truck. Both were sitting in the back of the truck's cabin, and I could easily hear their conversation. A cow was suddenly visible at a distance and caught his son's attention. Osvaldo told the boy: "Your papa is going to have some cattle one day, be a fazendeiro, and make you rich and proud."

As a technician, Osvaldo was one of the lowest-paid field workers in IMA. His status reflected a hierarchy that is ongoing in the organization. As with any other organization in the country, staff distribution in IMA mirrors Brazilian class structures. The division of labor is structured according to a scale of university degrees, research abilities, and tasks. This means that those with postgraduate degrees have more responsibilities, higher wages, and more recognition than lesser-educated staff members. Most of those in such positions are originally from southern Brazil or the United States. Because of its status as an independent research institute, IMA's distribution of positions and responsibilities is clearly defined in organizational diagrams, though the delineations exist only in theory. In reality, there are so many overlaps and flexible structures that projects flow in very different ways. Thus, many decisions within the organization were made in meetings, although a vertical decision-making structure exists. The diagrams, then, are useful for the organization's management, structural projects, governing body, and funding applications. In reality, however, decisions appeared to be centered on a small core of the four founding members, with an extended group of a few researchers for general discussions. The small core of founding members shared a similar profile, combining considerable academic output and proven fund-raising skills in their own projects.

Scientists at IMA carry out a type of applied science. Perhaps more aptly, their work could be described as a collective construction of engaged science, which it achieves through a coproduction of knowledge, or "thinking of natural and social orders as being produced together" (Jasanoff 2004, 2). In trying to define the problems that the Amazon region faces, IMA's scientists are seeking to shape policies. Jasanoff argues that "co-production is shorthand for the proposition that the ways in which we know and represent the world (both nature and society) are inseparable from the ways in which we choose to live in it" (Jasanoff 2004, 3). As an institution, IMA is modeled on the Woods Hole Research Centre (WHRC), a private nonprofit organization based in Massachusetts with a mandate to "advance scientific discovery and seek science-based solutions for the world's environmental and economic challenges through research and education on forests, soils, air, and water" (WHRC 2013). Most senior scientists at IMA have had some type of affiliation with WHRC, as have several of the junior scientists. Furthermore, many of IMA's scientific staff have been visiting fellows at WHRC. There have also been several collaborative projects between the two organizations. This partnership helped legitimate IMA's efforts in Brazil from early on.

IMA regularly carries out numerous projects, which I classify in three different categories: (1) community-based, (2) scientific, and (3) hybrid collaborative. All include scientific research. The difference between the projects is in how the research is applied. In community-based projects, studies are directly linked to people's livelihoods. One example is the Várzea Project, which is dedicated to fisheries management in riverine communities. Building on their studies of river fish and their behavior, IMA's scientists proposed adjustments to local fishers' practices to avoid depleting fish stocks. IMA's scientists were able to incorporate their findings into community practices without having to use any potentially baffling scientific jargon. Instead, they explained their findings and made their proposals in such a way that stakeholders were able to achieve a thorough understanding of the local river ecosystems and their dynamics. These changes have resulted in sustainable yields for fishers and their families and have provided clear results that are easily comparable with other areas along the same river. The second type of project is dedicated to collecting data from experiments and measurements in order to understand different relations among elements in the forest ecosystems. One example is IMA's aforementioned study of the effects of fire in the forest: IMA set fire in a controlled fashion to an area in the Amazon forest to better understand the plants' resilience or dying point. The results of this study will be used on other projects that assist local communities in dealing with fires, which are common throughout the region. The third type of project relates to IMA's collaborative enterprises, usually with other NGOs and sometimes with government offices, through which IMA seeks to orchestrate efforts to balance the welfare of local populations in the forest with different modes of repair for ecosystems that have been exploited and thus damaged by humans. An example of this is the aforementioned campaign to salvage springs that supply the Xingu River. This collaborative project is composed of a vast array of activities and involves local populations—including schoolchildren—and other institutional and organizational actors.

All of IMA's projects follow a formula that seeks the applicability of scientific knowledge to provide solutions to problems or to improve situations in the Amazon region. The formula can be synthesized as follows: (1) develop knowledge about a problem or situation by defining what evidence is relevant and measurable to define it; (2) seek to evaluate how relevant such knowledge is to local communities in their daily experience in the Amazon forest, and make changes if necessary to ensure a more inclusive exercise; and (3) use the resulting analyses to motivate individuals and policy makers to modify behaviors and policies in order to achieve a potential future. It is widely recognized that scientific facts "have interpretive flexibility" (Bijker, Bal, and Hendricks 2009, 2), which means that advocates can shape the data they collect in their analyses. The political process of negotiation that is carried out with such scientific knowledge is therefore crucial to the final interpretation that will inform the policy process. This means that the definition of problems and the potential solutions sought both require the "mobilisation of knowledge," that Delvaux and Manez explain is "the defining of problems and the fabrication of ideas" for public action (Delvaux and Manez 2008, 63).

As the most prevalent type of polity around the world, the state is responsible for policies that directly affect the lives of millions of people and the management of vast territories. The result of such accumulation of power can be a "tunnel vision" that privileges a few aspects of collective life—like economic activity and the building of infrastructure—over others—like the protection of the environment (Scott 1998, 11). In the case of the state, the most common priority—placed above any other consideration—is economic growth. In a meeting of environmental organizations in Manaus, one activist despaired that "Brazil's main scientific agencies are devoted to helping large agribusinesses instead of seeking better ways to balance [the] needs of the majority of the population." If the state's production of scientific knowledge is used for the purposes of economic expansion of the Amazon frontier, socioenvironmental advocates seemed to imply, then independent research will shed light on the implications of such a state of affairs. This is the type of legitimacy sought by organizations like IMA. At the Manaus meeting, Charles Clement, a specialist in plant genetic resources of the Brazilian government, complained about the lack of interest of the Brazilian government in funding research that would improve the understanding of the Amazon's vast and diverse territory. Out of 15,158 research groups that existed in the country in 2002, he said, only 590 worked in the Amazon region (only 3.9 percent). The gathering was the 2004 annual event of environmental NGOs funded by the United States Agency for International Development (USAID). His overall point was that many of the treasures of the Amazon forest are systematically ignored by an official policy that subsidizes research on commercial crops such as soy and cotton but does not explore the diversity of fruits, seeds, plants, and animals that live in the forest. He added that out of the 6,843 students who finished their doctoral degrees in Brazil in 2002, only thirty-eight carried out their graduate studies in the Amazon region.

In such a context, the regional relevance of IMA is evident. As an independent research institution, it carries out its scientific research in conjunction with local, national, and foreign universities. In many cases, links are to specific courses taught by its staff. In others, they undertake collaborative projects with individuals based elsewhere. Many of its staff members teach in the region. At the time of my fieldwork, IMA frequently hosted groups of students or invited one or two to be trainees. For its founders and leaders, however, scientific research is but a tool—a means to an end. It is complemented by two other activities: outreach and education. This process is in turn used to influence environmental policy making. This means that IMA both studies and works with small-scale producers while also carrying out investigations on large-scale scenarios. "I don't think there is virtually any other organization that works at the range of scales that IMA works at," Tony told me. "You've got [Pedro] doing global carbon international treaties, and we've got Projeto Tapajós, Várzea, Maflops, on the other extreme, where we are working with communities of grassroots producers on developing actual management in a marketing system."

Nathan linked this uniqueness to IMA's particular understanding of science—one that incorporates political strategy. He differentiated this from two other types of understandings, both of which he deemed partial and unhelpful: one focused on technology development, and the other focused on criticizing the development model. To illustrate his opinion, he provided an example of each type in the case of the Brazilian Amazon. "There's a tendency to think that scientists should be working on technology development," he told me, and went on to explain how because of the soy boom in Brazil, governmental scientific research institutes were largely dedicated to improving the quality of soybeans. The second type he found fault with "is much more antagonistic in that it's basically calling attention to mistakes and errors and damages associated with the development model. . . . Basically, the question is the paradigm." His favorite, though, is the third type, which "is done with the political strategy" that can be used "for sustainable development." That is why, he argues, this type is "the most potentially powerful to change the course of events, and yet, it's the hardest to achieve. And I think that you need to have watchdog scientists, you need to have technocratic scientists and politically relevant scientists, fostering your development model. I say we are missing out right now because huge amounts of money are going into technology, into very inefficient federal research agencies; quite a bit of money [is] coming in from international sources to watchdog [research]; and less is going into these politically relevant themes and regions."

Adrienne was of a similar opinion. In other conversations, she would explain a similar position to Nathan's from her own perspective. Once, she described to me a discussion she had with one of her lecturers in Boston while studying for her master's degree. She was complaining to him about foreign scientists who go to the Amazon, do their research, and then return to their countries without leaving anything beneficial behind: "He said, 'I know what you mean: that I do it and leave and that's it; but I believe I am doing my part because I am related to other research institutions that in turn have links to policy makers,'" to which Adrienne concluded in a condemning tone: "But if you are publishing without context then you did the research for yourself." Her words showed an ongoing concern that she had for political issues. This was also quite evident in the many meetings in which I saw her dealing with the leaders of social movements or with peasants as well as with other NGO personnel and government officials. She had a way of speaking that placed her own research in a wider context. She mixed a thorough analysis of her findings with an informed opinion of social concerns of the area, and she would spice everything up with jokes or comments about some of those present in the audience. She was well liked, and it was evident that her charismatic style for tackling controversial issues aligned her with IMA's founding leaders.

Adrienne's real edge among activists and social movements, however, was that she was local. She was not from a privileged middle-class background like most other researchers. She was originally from Belém and had earned her place in IMA. She was studying geography at the UFPA when she started working as an IMA trainee. She made a proposal for a study of the soil and received funding to explore it further. "I basically put together a database of soil types in Amazonia," she recalled in her office in Belém. "It was funny because I had never seen a computer before, but the person who chose me saw potential in me and it worked out." After her period as a trainee, she was offered her first job: assistant to one of the associate researchers. Then she spotted a series of "fire scars" in the Paragominas area and suggested a way of mapping fires that was well-received in IMA. "[Nathan] liked the idea and he gave me money, a car, and allowed me to take as long as I needed for the project," she recalled. "I developed a method to map fire scars in Paragominas. I interviewed almost all the big landowners within thirty kilometers of the city. It was a complete mapping on which I spent a year. That was the time when El Niño was very strong and the World Bank wanted a study that could give some answers about fire, and it was what I had just done. So they invited [Nathan] and me, and I coordinated the World Bank project to do the same in another four areas in Amazonia." This was back in 1996. She had just finished her undergraduate studies, and this project was written first as a report for the World Bank and then as a book. Then in 1998, she did a master's degree in remote sensing at Boston University, again with a project on fire. Her success within IMA, however, did not render her unapproachable to farmers or activists. In several meetings where I saw her coordinating groups, she joked around and was able to explain complex issues using everyday language.

Scientific research of ecosystems has become common currency in policy circles regarding the Brazilian Amazon. There was a time when government decisions that entailed the large-scale alterations of landscapes and social arrangements (e.g., road building or colonization) responded to nationalist or economic priorities. Nowadays, scientific evidence is used as a currency by the Brazilian government in its claims to design policies that serve a double purpose: to deal with accumulated conflicts (e.g., those derived from land tenure and soy expansion); and to reduce the pace of deforestation (to comply with international concerns over climate change and loss of biodiversity). As I pointed out earlier, the Brazilian government privileges economic growth over environmental concerns, in spite of some declarations by politicians or government publications that might suggest otherwise. Among the federal ministries, that of Environment is the least influential. Many activists, NGO personnel, and workers from international foundations repeated the same suspicions, which were confirmed to me by Manuela, who worked within the Environmental Ministry as overseer of all issues relating to the Amazon region. I met Manuela at the meeting regarding the paving of highway BR-163 in Alter do Chão. After a full day of activities, we talked in the restaurant of the hotel where she had earlier presented her keynote speech for the aforementioned meeting. I started out by asking about her past experience in civil society organizations. She told me that once President Lula da Silva reached office, numerous members of civil society organizations decided to work in the ministry, including herself. His left-wing government compelled activists who were working on many different issues to join a government that many political analysts considered to be the first leftist regime in Brazil's history. Before becoming a bureaucrat, she had led an environmental NGO in Manaus for many years. "In the Amazon region, all those belonging to the 'movement' decided amongst ourselves who would join the government and who would stay in their organizations. It was a strategic decision," she told me, adding that the "movement" still suffered from the migration of experienced staff from their organizations to the government. "But," she said with a smile, "long-term goals are worth the price. If we wanted to change policies, this is appropriate. Besides, most organizations were created not to perpetuate themselves, but rather to help define governmental policies." In her eyes, furthermore, policy making requires the participation of NGOs and other forms of organized individuals. The ideal situation was to have institutions such as IMA and others that produced high-quality scientific research feed the loop of policy making and governance. "These organizations have a triple advantage over universities," Manuela told me, after speaking highly of the scientist-advocates I had come to know within IMA. She elaborated: "They carry out research projects that are clearly engaged with problems and for which they try to find solutions; they benefit from more flexibility in research agendas; and they have more capacity to find alternative sources of funding for their projects."

One of IMA's key partner NGOs is the Socioenvironmental Institute (Instituto Socioambiental, ISA). With its prioritizing of indigenous populations,

ISA seeks to raise awareness of cultural diversity in a similar fashion to what the scientific community did with ecosystems: by establishing the concept of biodiversity. As Takacs showed, the emergence of the concept of biodiversity followed a route through which biologists sought to "forge a new ethic" (Takacs 1996, 9). As a concept, it was generally adopted in the 1980s within policy circles, perhaps influenced by the expanding environmental movement. According to Takacs, the concept "encompasses the multiplicity of scientists' factual, political, and emotional arguments in defense of nature, while simultaneously appearing as a purely scientific, objective entity" (Takacs 1996, 99). In framing biological processes through the concept of biodiversity, ISA shows that it values various aspects of ecosystems and life forms. Similarly, ISA seeks to bring attention to the variety of indigenous populations in the Amazon and their interactions with the forest. Among ISA's publications, for example, is a massive volume that analyzes areas dedicated to biodiversity conservation and to indigenous populations (Ricardo 2004).

ISA and IMA are two of the most visible NGOs in Brazil with a socioenvironmental agenda. As strategic partners, they collaborate on projects where each organization's strengths make valuable contributions. Both handle large budgets to carry out numerous projects. Both use scientific research as a central element of their campaigns. Whenever he would explain IMA's work to policy makers or international foundations, Nathan was adamant to convey the sense of responsibility IMA researchers have. On the one hand, he said, they need to carry out their scientific research with a high degree of quality. On the other, he referred to the need of their endeavor as a "survival enterprise," for "showing results in the world" that would make a difference. It is the difference between universities and scientific NGOs, he said; in the latter, "it's all soft money, that means you have to raise money, and that means that you have a built-in incentive to do relevant stuff. If you are just doing research that is more theoretical it is very difficult to get funded." He insisted that IMA's agenda is about "leveling the playing field: giving social movements, grassroots movements, 'fire and tools,' the information in maps and projections that we give, to bargain on an equal basis." Nathan did not seem to mind that his comment regarding "fire and tools" implied that their job was to "modernize" local communities. The view that IMA is a modernizing force is common among some of its members, which reinforces a sort of "capacity hierarchy" and "distinction" between themselves and the stakeholders with whom they work.

⁓

In this chapter, I demonstrated the role of knowledge production as a purposeful practice of NGOs like IMA and of the networks in which they take part. Because knowledge is somewhat malleable, its uses for political means are factional. Within NGOs, activists and advocates seek to influence government policy making in order to modify what those involved in the networks deem problematic. In this sense, IMA is on a mission. It uses its own scientific inquiries as a means to an end: to change what is taking place in the Amazon. The change sought is of hearts and minds. The IMA hopes that concomitant with this change will be a modification of both government policies and the everyday practices of the region's inhabitants. By offering an informed view of a potential future with a better balance between the protection of the biome and the well-being of the region's population, IMA's scientists appear to be aiming at constructing hope. They do this through an agenda of step-by-step actions required to improve situations. Would this not be equivalent to an exercise in the building of morale? Or of nationalism? Would this use of scientific research by organizations with a socioenvironmental agenda imply an effort to provide a wider, more inclusive vision of a potential biosociety?[2] In recent decades, the distinction between nature and culture has been dismantled along with the dream of human control of the environment. Taken in combination, these two factors have driven individuals to question the assumptions that had been generated within a narrative of modernity (Latour 1993). IMA and other NGOs ground their work in the conviction that scientific data will not achieve the change sought, but that the data are an essential part of the effort. In this sense, these organizations are not serving a modernizing agenda that tries to eliminate ancestral myths and traditions, but rather are seeking to incorporate these into a collective effort that includes others' beliefs. As Lévi-Strauss pointed out, "my feeling is that modern science is not at all moving away from these lost things [mythical thought], but that more and more it is attempting to reintegrate them in the field of scientific explanation" (Lévi-Strauss 1978, 3–4). NGOs who put socioenvironmentalism into practice combine various traditions and pepper them with scientific information in order to garner the legitimacy needed to influence policy makers. As has been evident in the last few years, however, it seems that such efforts have so far not been enough to modify decades of extractivism and developmentalism. But as practices of advocacy networks, they are alternative performative politics. The following part seeks to build on reflections about aspirations, from those of the individuals who take part in advocacy networks, to those that such networks deal with in their negotiations of vernacular and cosmopolitan principles.

---

2. This term has been used to indicate a leap in technological advances. Here, I use it to include the biome, in tune with Ingold's idea of humans dwelling as organisms in an ecosystem (Ingold 2000a, 2011).

PART FOUR

Informed Aspirations

7

Moral Entanglements

How can we work with all the interests and groups to reduce differences and conflicts and increase the [potential for] consensus on a vision for a landscape and the sustainable use of its resources?

    —João, at a meeting of NGOs and social movements in
    Manaus convened by USAID

THE VIEW OF the Amazon River from the upper floors of the luxury hotel in Manaus was intriguing. The moist air after the rain produced a veil of fog that was difficult to distinguish from the river. Inside the hotel, dozens of activists and advocates from numerous NGOs, farmers' unions, and other social movements were taking part in a conference convened by the Brazilian chapter of the United States Agency for International Development (USAID). There were some keynote lectures and a few parallel workshops and events. I was attending alongside a large group of Instituto do Meio Ambiente da Amazônia (IMA) researchers. In one of the workshops, chaired by Marcelo from Instituto Socioambiental (ISA), discussions were guided by the question João posed: "How can we work with all the interests and groups to reduce differences and conflicts and increase the [potential for] consensus on a vision for a landscape and the sustainable use of its resources?" After a couple of interventions that emphasized difficulties or coincidences between participating NGOs and social movements, Marcelo decided to state his ideas clearly: "I do not believe there is such a thing as a neutral arena. Each [institutional] actor is an actor because it has its own interests. What is important is to have clarity in one's own. The ideal mediator should be the government, but the government is schizophrenic and has its own agenda. What is important is to start from our common goals." He then went on to explain some of the difficulties that advocacy networks face: "Firms do not want to talk to NGOs, but also NGOs have difficulties in talking with firms. There are prejudices against each other. If on a table two yell at each other 'you are a stingy egoist' and the other 'and you an intransigent radical'—" Somebody joked: "Both would be saying the truth!" and all present laughed. "Precisely," agreed Marcelo, but added, "but both are intolerant with each other and are not willing to listen to each other." The challenge, he insisted, was to allow for dialogues through which agreements may be reached. Participants constantly interrupted each other's comments with jokes or comments of their own, but these served to lighten the mood that was evidently tense because of the contrast in participating groups attending the conference.

One of the questions that has guided my inquiries in activist networks is: how is it that such large numbers of people, many of whom would not relate to one another in different circumstances, come together to work for a purpose they often jointly define? The question is easier to apply to individual activist groups because the convergence is more logical: each group is usually bound by specific goals or ties (of friendship, ideology). Depending on its members' aims, the group can be either small and informal or large and bureaucratic. Once a group decides to join efforts with other groups, the contrasting work cultures of each group can be problematic for the network. Sometimes, it works. In both field sites—Spain and Brazil—I encountered groups of varying sizes, complexities, and resources. In both, I witnessed networks whose members were compelled to continue collaboration despite challenges. I attributed this to the emerging moral entanglements among their members.

This chapter makes the case that "moral entanglement" is a useful category to appraise the challenges advocacy networks face. A moral entanglement is here defined as a correspondence in certain beliefs of what is right and wrong between otherwise contrasting group agendas or frameworks. An example in the Brazilian Amazon was the pursuit to reduce the destructive force of fires in the forest. Numerous NGOs, peasant unions, government offices, and international agencies coincided in the need to do so. Such a commingling, however, came with its fair share of growing pains. Collaboration was nevertheless the result of a consensus reached through difficult negotiations. The groups avoided an outright condemnation of the practice of burning a piece of forest to clear it, instead collaborating to pursue ways to avoid such fire getting out of control. This was a pursuit all groups felt they could engage in without feeling as though they had abandoned their core beliefs. Scientists still thought it would be better to avoid fire altogether, but they came to understand that using machinery to clear a piece of land of large trees was not practical deep in the forest. At the same time, peasants felt their long-used practice was being recognized, and they felt empowered even while

Figure 7.1. Farmer checks that a controlled forest fire stays within the area that it is meant to in a practical workshop carried out by the Instituto do Meio Ambiente da Amazônia (IMA). (Raúl Acosta)

they enjoyed learning new techniques to keep fires under control (figure 7.1). With decisions like this one, groups participating in advocacy networks negotiate both solutions to identified problems and the sense of collective endeavor, which is crucial for their civil becoming.

The types of associations analyzed in this book are actually moral networks; that is, they seek to collaboratively promote a moral compass that they can use to navigate the issues in which they are involved. Just how fair and equal such coproduction of the moral compass is, however, requires examination. There are several factors that tilt the balance of what is defined by whom (for example, rhetorical capacities, access to various kinds of knowledge, resources), just as there is often a certain general framework that, from the start, limits the terms of discussions and possibilities (for example, regarding environmental protection in the Amazon). The full effect of discipline regarding environmental issues falls within a framework of environmentality (Agrawal 2005). In his acclaimed book *Environmentality: Technologies of Government and the Making of Subjects*, Agrawal builds on existing analyses of environmental politics in political ecology, common property, and environmental feminism to present the concept of environmentality (Agrawal 2005, 5). His framework considers political decisions on environmental affairs and crucially the shaping of subjectivities through changing institutional arrangements and discourses. In his words, environmentality "constitutes a way to think about environmental politics. It attends carefully to a) the formation of new expert knowledges, b) the nature of power that is at the root of efforts to regulate social practice, c) the type of institutions and regulatory practices that exist in a mutually productive relationship with social and ecological practices, and which can be seen as the historical expressions of contingent political relationships, and d) the conducts that regulations seek to change, and which go hand in hand with the processes of self-formation and struggles between expert-authority based regulation and situated practices" (Agrawal 2005, 324). This means that even modes of resistance or alternative practices (Cepek 2011) work within a framework of competing forms of knowledge and aspirations to manage the environment according to emerging subjectivities.

The cases I observed put in practice an approach to environmental politics that fits what Agrawal calls *environmentality*. This does not mean, however, that activists fully achieve their goals. For example, in their efforts to control forest fires in the Amazon, NGOs are aware that it is virtually impossible to do so all over the region. In seeking to clean the springs of the Xingu River, they know their work will be inevitably incomplete. I asked several members of IMA if they were not disheartened by the sheer scale of problems that they faced in the Amazon. Although their replies differed, they shared an acknowledgment of the problems, and a positive outlook of their own work. Adrienne, for example, trusted that with time their endeavors would pay off: "Our work is slow, because more than just research we engage with social changes that require time." Difficulties were not limited to the stakeholders with whom they worked, but rather included the collaborative work with other NGOs and participating groups in the networks they were part of. In carrying out campaigns, programs, and activities, IMA engaged in a series of conflicts and stalemated negotiations within other groups. Nevertheless, even with their inequalities, the networks I observed managed to collectively shape a sense of the good—an aim that all participants would be able to showcase as a positive practice to strive for—through their friction-laden, long-term, social interactions. Through such synergies, I argue in this chapter, the groups involved not only worked to establish parameters for adequate behavior, but also legitimated their role as "civil actors." Looking outward from these specific instances of fieldwork, I suggest that by reaching agreements on various topics, advocacy networks are shaping a grand exercise of ethical imagination with a transnational reach. The value of such work is that it has to be done through cycles of negotiation in various scales and contexts. For each setting, local historical events and their repercussions in current configurations must be included in negotiations.

The sets of networks studied in this book show contrasting results in this respect. The networks in which IMA is involved have managed to establish mechanisms to negotiate among stakeholders, using various types of knowledge. In so doing, they are seeking an ethical standard of negotiation and commitment. The Mediterranean Social Forum (Fòrum Social Mediterrani, FSMed) sought to establish a working network in the region to foster periodic meetings such as the one I observed in 2005, but this effort was cut short. The many conflicts and problems within the preparatory process before the first meeting, and perhaps even the contradictions between ultimate goals and methods used to reach them, proved to be too much for a continuous commitment in the framework of the FSMed. But perhaps that in itself is a seed of civil becoming.

This chapter is therefore an account of the agonistic basis of the process through which advocacy networks become morally entangled. The practice-orientation of advocacy networks means that moral principles are carved out of practices. In this sense, it is considered here that activists and advocates first seek to establish good practices as ethical, before reifying the underlying principles as moral. What follows, therefore, is first an exploration of the concepts of ethics and the ethical imagination. The chapter then delves into a consideration of how the resulting moral entanglements serve as a legitimation of the ongoing construction of civil becomings among participants in the networks. For these processes, NGOs and other groups draw from the religious heritage of many advocacy efforts to practice an idea of the future as a path toward a secular salvation (Mosknes and Melin 2013). In doing so, advocacy networks practice secular ideals as transcendental moral principles.

Ethical Liaisons

When I joined the organizational meetings of the FSMed in May 2004, they were being held at the offices of the Fons Català de Cooperació al Desenvolupament (Catalan Fund for Development Cooperation). Alicia, who headed the Fund, took part in the meetings with a clear interest in getting as many southern activists as possible to attend. The Fund had given a large grant to the FSMed to ensure that some activists from the south could attend the preparatory meetings that were being held in different countries around the

Mediterranean. "We believe that by providing financial assistance for travel, these groups will benefit in taking part in the forum, in shaping the forum to be what they need and want," she once told me in her office. Later that year, the Technical Secretariat hired two people to assist with organizational issues. One of their tasks was helping activists from outside of Europe apply for visas for their attendance at the four-day FSMed event in June of the following year with the relevant Spanish government authorities. This way, individual activists would not need to apply by themselves through a consulate; rather, all applications would be processed directly with the contacts in the foreign ministry that FSMed organizers had established. Catalan activists insisted it was an essential part of any event within the World Social Forum process to try to level the playing field as much as possible to ensure the participation of groups that would otherwise not be able to participate. Such principles of assistance, however, did not always flow smoothly. For the preparatory meetings, for example, it was not always possible for the same activists from southern countries to attend. For the visa applications, the information that was requested from applicants took longer to arrive than was originally envisioned, thus straining the team's capacity to follow up each case individually.

No one involved in the preparatory process was in doubt that helping those groups that most needed the aforementioned assistance was an ethical principle. Assistance of this nature was thus an example of an ethic, or a principle, put into practice. Although not in such a formal manner as professional associations do with their "ethical guidelines," activists organizing the FSMed still had a clear idea of what good practice meant. Such clarity had been gained through the combined experience of participating organizations. "Each of the participating groups in the Technical Secretariat already collaborates with some groups in the southern Mediterranean," Javier told me over a *caña* (beer) after one of the meetings. He added, "In a way, this [the FSMed] is about bringing some of our networks together." In doing so, they were also combining to some extent the ethical dimensions of their interactions. For some organizations, these appeared to follow a patron-client model, but for others it was less hierarchical. The whole idea of the FSMed was to enable interactions between groups that would have difficulties getting together without such an event.

Ethical liaisons, nevertheless, went beyond financial or bureaucratic assistance. They were part and parcel of the links connecting all groups involved. Perhaps two spheres of ethical entanglements stood out: the constituent practices *of* their activism, and the coproduced enacted practices *through* their activism. While the examples of financial or bureaucratic assistance belong to the former, the definition of controlling fires in the Amazon belongs to the latter. The underlying assumption activists and advocates share is their belief in being involved in "doing good" (Fisher 1997). As Javier once told me, though, it is through their participation in networks that they can place their own plights within a bigger picture. Such visions, though, had very specific repercussions in terms of activists' choices and actions that reflect the experience of networks more broadly. In acting together, activists are often confronted with different ways of addressing similar problems. This stresses the processual character of their ethical practices.

Anthropological approaches to morality and ethics have faced a dilemma similar to the one that has plagued numerous debates on the concept of culture. While culture has been recently demystified as static and is now considered as a more fluid set of practices, beliefs, and references (Kuper 1999; Boggs 2004), its invocation in other academic disciplines and in public debates often provides a sense of continuity and self-containment. A recurrent assumption when speaking of contrasting moralities is that there is only one morality—or one predominant morality—for each particular society. However, this is a contentious point, as Zigon notes: "For anyone who has done explicit anthropological research on moral belief and practice, this is clearly not the case" (Zigon 2008, 12–13). MacIntyre argues that "moral concepts change as social life changes" (MacIntyre 1998, 1–2). This means that it is through interactions and negotiations that ethical principles are established. Zigon argues along these lines by insisting that our way of life is not simply transmitted from one generation to the next: "Indeed, if it were true that all of our judgments, knowledge, beliefs, perceptions, and practices were culturally learned, that would more or less render us cultural automatons who were trapped within what Hallpike calls the tyranny of culture" (Zigon 2008, 15). And just as it was for most of the twentieth-century anthropologists' for whom cultural relativism seemed to portray a neatly differentiated set of cultural milieus, moral relativism entails a vision of bounded societies that simply does not exist.

The opposite extreme is that of moral universals that presume that principles are everywhere applicable. Both positions represent caricatures of actual phenomena, or, for some, ideal types. If morality is only shaped through interactions, then systematic efforts to influence those interactions will have an effect on moral principles. Such is the case with religion, which provides a basis for beliefs and practices that are disseminated to a population, along with a series of promoted principles. When some religions spread farther than where they originated through proselytizing and missionary work, they provoked a series of cultural negotiations that allowed for those religions' principles to be adopted far from their origins. In some cases, these negotiations involved the incorporation of vernacular rites, ceremonies, symbols, or concepts within the expanding religion, which often resulted in hybrid forms.

These, in turn, with time and changing practices, were either absorbed by the invaded culture or became separated from the founding religion altogether. This process mirrors that of the moral frameworks that evolve through contact with contrasting principles and changing attitudes to what is considered appropriate within specific social milieus.

Advocacy networks seek to provide a moral compass for debates by highlighting issues that require attention. They may, for example, take up cases of human rights violations, environmental damage, or gender inequality. It is often the case that some advocacy networks apply a framework with universal principles without calling it as such. In doing so, they pose a challenge to a long-standing anthropological position. Over the last hundred years, anthropologists have tended to avoid universals in order to steer clear of Western ethnocentrism (because, in the end, who decides what universals should be?). While debates about the permeable character of culture have reached far and wide, however, "many anthropologists today still share many of the assumptions that underlie compartmentalized cultures, which are bounded and cut off from other cultures" (Zigon 2008, 144). Anthropology thus holds on to its dual heritage from the Enlightenment and Romanticism. On the one hand, it purports the certainty that humanity can be studied scientifically in its different dimensions (biological, cultural, linguistic, and social). On the other hand, it values local conventions by insisting that people have distinctive features according to their own sense of collective self. This tension implies an inherent difficulty in any effort to understand peoples from a distance. If we take seriously anthropological debates about culture and their implications over the last few decades—basically, that there is no static culture—we can better understand the negotiations that take place among populations and individuals from far and wide. These deliberations tend to leave marks on all those participating. For Zigon, "morality is better thought of as continuous dialogical process during which persons are in constant interaction with their world and the persons in that world, rather than as a set category of beliefs from which one picks appropriate responses according to particular situations" (Zigon 2008, 155).

If morality shapes what is appropriate, then ethics is more of an "embodied disposition," or a "reflective and reflexive" moment about the better way "of being in the world" (Zigon 2008, 165). This ethical process takes place individually, through the self, and collectively, as an exercise to shape an ethos within a collective. Examples of such collectives are professional associations, families, or religious congregations. Some of the largest examples are guided by the notion that belonging to a nation, in which members ostensibly share certain characteristics either by default or by choice, can stir up a sense of belonging and kinship in those members. With help from artists and writers, as well as media and communication technologies, such imagined communities provide a narrative through which certain traits are easily ascribable to its members, often promoted and exploited by politicians and other power figures (Anderson 2006). Of course, a single individual belongs to a range of collectives at the same time (Strathern 1990). This implies that one person could belong to various collectives, each of which may promote contrasting behavior as ethical. What takes place, therefore, is a negotiation within each individual, to decide for herself on a desired appropriate behavior. Such negotiations also take place within each of the collectives; as its members are never blank slates, they may arrive with many references of various codes of conduct.

Because states in their modern form seek to control social life in its many dimensions, its norms and laws tend to be framed as ethical standards. In reality, however, the distance that exists between professional politicians and the population makes their aspiration to gain total control simply unrealistic. While laws and norms do respond to negotiations within the collective that forms the state, they nevertheless primarily take shape among institutional interest groups that have come to dominate the realm of politics. In order to understand the subtle exchanges taking place between powerful actors, Mouffe's distinction between politics and the political is helpful. While politics is the set of institutions and practices through which a social order is created, the political is the dimension of antagonism constitutive of human societies (Mouffe 2005, 9). I would add that such antagonism is not simply any type of antagonism, but one whose raison d'être is what is shared, or public. Among any group of people, there will be disagreements about how to deal with an issue that concerns all those involved in the collective. This will be the case especially when the members have contrasting references for what is morally appropriate. In a society with only one religion and without much social stratification, chances are that opinions will be similar. Such societies, however, are few and far between. What is more usual is to have a wide variety of influences and references crisscrossing any social assemblage.

The political, as the space shared within a social collective where negotiations regarding difference take place, is also where people enact the "ethical imagination." For Moore, ethical imagination represents "the way in which technologies of the self, forms of subjectification and imagined relations with others lead to novel ways of approaching social transformation" (Moore 2011, 15). It is through imagination that the possible takes shape. Imagination may motivate actors to adopt certain practices that seem appropriate to try to reach a goal. It may also help actors compare their situation with similar but alien ones, which may in turn assist them in achieving their aims. In any case, what is brought about is a "refiguring of self-other relations" that "provides for a series of complex links between objectifications, stylization and agency" (Moore 2011, 18). Only by imagining potential outcomes can one decide to establish them as aims.

What ethics entails therefore is a management of self in relation to others (Moore 2009). Individualism does not mean a reduction in the inherent codependence of humans (Moore 2011, 203). In acknowledging such codependence and using such knowledge as a basis for grand political horizons, clearer considerations regarding the balance between individuality and mutual reliance may develop. This equilibrium must nevertheless build on recent reflections on culture. It should be clear by now that we cannot conceive of human groups as holding perennial cultural configuration (Eriksen 1993b). Conceiving of groups in this way would imply that we understand the process of change that occurs among us. As Bloch says, "there are no lasting boundaries within the human conversation; change is rapid and different in every case because the interactions between people are so numerous, so volatile and so extraordinarily complex and because every individual is combining different configurations of factors in the context of unique situations" (Bloch 2012, 34).

While institutions seek to phrase their efforts to advocate for moral orders in language that implies its members are following enduring principles, the actual process is more haphazard. The groups gathered in the FSMed Technical Secretariat, for example, seemed to share some principles but were clearly divided on others. The difference I identified between radicals and reformers was due not only to each group's core beliefs and practices, but also to their members' life histories. María, for example, who became the most visible proponent of the reformers' side, was a middle-class native Catalan who was comfortable working for an NGO and carrying out diplomatic relations with other NGOs and government officials. Javier, however, whose voice among the radicals was respected and sought after, was a working-class Andalusian immigrant to Catalonia. Although there was a common language of justice and rights among all participants, the differences between both camps often generated frictions regarding contrasting takes on nuanced interpretations of what justice and rights meant. Each of the two camps also slowly coalesced during the preparatory work of the FSMed, perhaps in a learning process where some of the participants identified with each other more than with the other group. And yet, despite their differences and conflicts, those in the FSMed Technical Secretariat managed to carry out the event that had brought them all together in the first place.

Time is a central element in any sociocultural negotiation of the sort portrayed in this volume. It is also one of the concepts that provoke those who espouse the principles of relativism to point out that there are numerous contrasting forms of perceiving or representing time. But when there is a need to reach agreements between a wide range of groups and ultimately with state governments, there is a requisite to engage in the standardized vision of time that has become prevalent around the world. In this sense, for example, indigenous groups managing their own reserves in the Brazilian Amazon need to engage in dialogues with NGOs and other groups who are considering the immediate past and imagining the future. It is what Bloch names "time travel": "We use time travel for all sorts of things, some more mundane than others, ranging from planning in making a list of what to buy in a supermarket to recalling past episodes of our lives, or the evoking of the past and the future in myth and poetry" (Bloch 2012, 108). In his view, such ability entails an exercise of imagination and is "at the very root of human social life" (Bloch 2012). It is one aspect of what Massumi calls the virtual (Massumi 2002). With this in mind, I present a reflection on the characteristics activists and advocates allocate to time: How do they imagine the future for which they can strive? What is their role in bringing it about?

Visions of Common Futures

At the end of a day-long meeting with leaders of movements based along the BR-163 highway (see chapter 4), Adrienne was providing the last presentation with a data projector and slides on a screen, trying to summarize the implications of what had been discussed thus far. Her initial slide was a photograph of the BR-163 in its state at the time: a dirt road. With a click of her finger, a gray surface with a yellow line in the middle of the road appeared, simulating what it would look like if it were paved. "Today we have seen what we all agree on: like, for example, that we all want the road to be paved," said Adrienne, and everyone agreed. Then, Adrienne went on to present a controversial example of how land could be redistributed in order to make the most of the characteristics of the terrain. Some of those present immediately complained, with one jumping up from his seat and saying loudly: "I cannot believe it!" He was a representative of a union of small farmers, and added: "This is equivalent to giving our lands away!" She asked everyone to calm down. "This is just an idea; it doesn't mean we are promoting it or that it is part of any plan," she explained. She asked everyone to allow her to finish her presentation and leave their comments for the end, which had been the procedure throughout the day. Her presentation moved on to other aspects, referring more to what could be the purpose of the network that was being formed. "There needs to be a clear message with the aim of our network, and that is what we are looking for," she explained. Once she finished, a round of comments followed in which the controversial slide figured prominently. Some of the IMA members present in the room explained the purpose of

Adrienne's exercise in imagination while still empathizing with the reactions it provoked. Other participants also took their turn to speak, arguing that such a rearrangement as Adrienne proposed was not a priority because it would cause more problems than solutions. A large part of the ongoing conflicts in the Brazilian Amazon forest are related to land tenure: overlapping claims exist for much of the land on which various government colonization programs have taken place for the last few decades. This explains the heated reaction of small landowners, who saw Adrienne's suggestion of redistribution of lands as creating yet another situation that would lead to potential conflict over land tenure.

What the scene described above brought to my attention was the ongoing pursuit by IMA and its partner NGOs to use maps and various modeling techniques to represent what the future might bring to the Amazon region. In several meetings I attended, IMA presenters introduced different versions of models that were generated with data taken from their scientific measurements. One example that summarized large amounts of their research data and analyses was published as a letter in *Nature* in 2006 (Soares-Filho et al. 2006). It compared two scenarios that IMA scientists and other collaborators considered would be the fate of the Amazon basin (including all countries where the forest reaches): business as usual, or with governance in place. In the letter, the authors claimed that conservation was threatened by the expansion of soy cultivation and cattle ranching, especially at a time when there was a push to open "all-weather highways"[1] into the region's core. They provided data to show the outcome of current trends of deforestation and argued that a more comprehensive strategy was necessary to contain the rate of deforestation.

Such explanations as the ones outlined in *Nature* are exercises of imagination. They use scientifically generated knowledge to invite government authorities to change their policies. But, they also entail an element of socialization—of sharing such information as widely as possible so that all stakeholders can understand the implications of their combined practices. What the networks in which IMA participates do is try to generate enough awareness among stakeholders and government officials to stimulate combinations of ethical behaviors on different scales. It is what I call "ethical layers." If members of each stakeholder group behave ethically regarding their main practice (e.g., farming) and promote ethical behavior among other stakeholder groups, then the push for change will be collective. What IMA and other groups appear to do is to seek a wide agreement about changes that are

---

1. Dirt roads in the Amazon region become extremely difficult to navigate during the rainy season, especially for heavy trucks.

necessary in the region. Such a process is necessarily slow, as it requires dialogues between groups that vary in size and profession, ranging from small groups of local farmers to large gatherings with policy makers and representatives of international agencies. It also requires trials of good practices in order for those involved to participate in the generation of the ethical principles. It may be easy to assume that because of the expertise of IMA's scientist-advocates, their expert opinions would be used by government officials as formulas to improve practices related to environmental protection. What I witnessed, however, was a more comprehensive work by IMA's personnel. Rather than relying strictly on their own knowledge to change the behaviors of local inhabitants, they guided locals through practices that incorporated other forms of knowledge as well, including those of locals themselves. Indeed, their analyses of the situations they were researching served only as an initial step in the entire process. This process was carried out at the same time as IMA entered into the negotiations with government policy makers that added political legitimacy to their scientific projects. Such was the case with projects for sustainable fisheries as well as for furniture making out of fallen trees in a remote area of the forest.

Advocacy networks generate an ethos from the interactions between their members. This ethos is partly related to the aims that bring the member groups together into a network, and partly related to the legacies of associationism that each grouping brings to the mix. Some authors argue that networks represent the ideal of horizontality in decision-making. Maeckelbergh, for example, argues that the global justice or alterglobalization movement is "a radical democratic alternative based on a decentralised network structure and principles of horizontality and diversity, and it intends to turn this alternative into reality through prefiguration and connectivity" (Maeckelbergh 2009, 38). Horizontality, she goes on to argue, "is something that one *does*, and if one acts like a horizontal, one is a horizontal" (Maeckelbergh 2009, 58). In her assessment, horizontality entails an attitude aligning with democratic principles in which consensus is sought among all participants. She explains that its opposite is verticality, or the concentration of decision-making among a few who thus situate themselves above the rest. For Riles, however, horizontality is often a myth. She argues that the illusion of horizontality can be dispelled in groups professing their dedication to it through such aspects as seating arrangements in meetings (Riles 2000). From what I witnessed in the FSMed organizing process, Riles has a legitimate point. Although activists stressed that anyone should feel free to contribute, debates were monopolized by leaders of the largest coalitions or organizations. In any meeting, those individuals who have more information or who better understand the nuances of discussions will have more legitimacy in the eyes of other participants than those who do not. Instead of considering if this fact discredits advocacy networks and their endeavors, I believe that what is essential is to understand the reverberations of action these networks produce.

When advocacy networks engage with the idea of "improving" a certain situation, they do so with an aim of changing the future, of producing an outcome that will be possible only through their intervention. This envisioned future predates the network already being present in each of the activist groups as well as in the individuals who form part of them. It is an agentic principle of acting, of making a difference. The crucial point about advocacy networks is that when a wide diversity of groups coincides in a common goal, their proposition is given greater legitimacy on account of the joint nature of their endeavor, even if their efforts thus far to achieve a common goal have been vastly different.

But what does this mean in terms of time? To begin to answer this, I draw on a recent debate between Maeckelbergh and Krøijer about radical left politics and linear time. In *The Will of the Many*, Maeckelbergh refers to radical left activists as working toward a future they envision by making sure there is no difference between the how and the what in the struggle: "In this sense, practising prefigurative politics means removing the temporal distinction between the struggle in the *present* towards a goal in the *future*; instead, the struggle and the goal, the real and the ideal, become one in the present" (Maeckelbergh 2009, 66–67). She argues that through such conflation of ends with means, the values of the envisioned society can be best put in practice. She uses the concept of "prefiguration" to frame this process. Krøijer points out that among the radical activists that she studied, such process did not have the linear character to which Maeckelbergh was referring. She therefore sought to develop a nonlinear argument about time that "highlights how time among activists has different temporal ontologies" (Krøijer 2015, 27). Drawing on Eduardo Viveiros de Castro's theorization of Amerindian perspectivism (Castro 1992, 1998, 2004), she advanced a perspectivist model of time in which "the future is not thought of as a point ahead in linear time, but as a coexisting bodily experience" (Krøijer 2015, 28). This means that through their practices, activists are already enacting the world they want to live in.

As part of her argument, Krøijer uses the concept of "political cosmology" to argue that what activists share is not an ideology but a cosmology, which she employs as a heuristic term to refer to "the logic of activists' perceptions of the world around them, the order of things in the world and the possibilities or radical change" (Krøijer 2015, 38). She argues that it is not so much through ideas and arguments that activists make their convictions known, but through actions such as self-organization direct democracy—which is what Graeber calls "small-a anarchists" (Graeber 2002, 72). Both

Maeckelbergh and Krøijer focus their analyses on radical left activists, and both refer to the World Social Forum and similar events as spaces for "reformist" groups; that is, these events differ from those of radical activists. To clarify the differences in terms of time perceptions, Krøijer argues that trade unions and NGOs are "oriented towards the near future, and use the social forum to discuss and launch more moderate advocacy campaigns or proposals for policy reform" (Krøijer 2015, 41). Radical left activists, on the other hand, seek out a radical change from the roots of capitalist society.

Albeit with clear differences due to Krøijer's and Maeckelbergh's choice of radical leftist groups, I find their reflections on using a temporal perspective as a guiding principle for enactments of activism productive. In the networks that are portrayed in this volume, there is a conception of time that, although not outright linear, does refer to aggregative effects of efforts as being conducive to a desired outcome. This is in line with the idea of "development" prevalent in international policy making and some scholarly circles. I would not call this outright linear because there seems to be an openness to a variety of potential outcomes, with a possibility of unwanted results. Many of the NGOs and organizations participating in the FSMed considered their roles as making contributions to "improve" unjust or unfair situations. But, such improvement was not represented by a single direction. Overall, the organizations participating in the FSMed generated a less radical atmosphere than those to which Maeckelbergh and Krøijer were referring—although some radical activists were welcome to join the FSMed's preparatory meetings if they accepted that the overall push would not be as radical.

The wing of the FSMed groups that I have called radical was clearly less radical than those portrayed by Krøijer and Maeckelbergh. They were not the sort of groups that use violence on the streets to make a point. Their radicalism was in their openness to criticizing government institutions or processes as well as some NGOs, whom they perceived to "be collaborating" with the government by rubber-stamping unfair processes or whitewashing unjust policies, thus perpetuating inequalities that they claimed to be working against. When the Forum Barcelona was taking place, radical members of the FSMed wanted to boycott any NGO that had taken part in it. This was the case until they heard about a situation that changed their opinion. Javier told me about it over a coffee in a small restaurant below his office one morning: "We have heard that the Catalan government threatened to cut some NGOs' funding if they did not take part in the Forum Barcelona." Suddenly, the radical members of the FSMed suddenly felt more empathy for the NGOs involved. The government had almost blackmailed them into participating, but they could not openly admit this for fear of their funding being cut.

In the Brazilian Amazon, the networks I observed also worked less radically that those Krøijer and Maeckelbergh describe. Part of my interest in looking for wide networks was precisely to understand how groups with very different backgrounds and aims could collaborate. I seek to analyze transnational advocacy networks as a result of historical processes in which various individuals aim to combine their efforts in order to seek stronger demands for changes that have more chances of being incorporated in state government policies. As motors for change, advocacy networks are actually not far from the aims of radical activists. A key motto of the alterglobalization movement, for example, is the title of John Holloway's book: *Change the World without Taking Power* (Holloway 2005). His anarchist viewpoint is widely shared among social movements and some academics, although distrusted by supporters of other ideological formulas. Its message, however, captures the imagination of activists seeking to avoid what is deemed to be the inevitable corruption that comes with power. Holloway's is therefore an ethical stance that was forged in the aftermath of historical events. Time and again, experience tells, revolutions have been followed by systems that tend to centralize power in one person or party (examples abound, such as the French, Mexican, or Russian revolutions). In this sense, Gandhi or Martin Luther King, with their calls for pacific struggle, have been examples to many organizations advocating rights or other issues. In essence, these peaceful protests or demands draw inspiration from Étienne de la Boétie, who is remembered for his work on early anarchism. In his youth in sixteenth-century France, La Boétie wrote an essay whose title has been translated as *The Discourse of Voluntary Servitude* (La Boétie and Bonnefon 2007). In it, he scrutinized why people allow themselves to be subjected to tyranny: "For if tyranny really rests on mass consent, then the obvious means for its overthrow is simply by mass withdrawal of that consent. The weight of tyranny would quickly and suddenly collapse under such a non-violent revolution" (La Boétie cited in Rothbard 2007, 9). His political thought was based on two premises, which Rothbard summarizes as follows: "the fact that all rule rests on the consent of the subject masses, and the great value of natural liberty" (Rothbard 2007, 9). La Boétie's ideas have been used in numerous pamphlets throughout history and have served as inspiration for political philosophy.

Although many of the groups who form part of advocacy networks would never consider themselves as activists or anarchists, their activities actually respond to the same principle. If they believed that government policies actually worked, they would not feel the need to organize demands to change them. If they considered certain populations as already possessing certain knowledge or education, then they would not find it necessary to organize courses or workshops. Numerous NGOs, for example, work with the assumption that their contribution is necessary for providing an ampler showcase of options than those available. In many cases, campaigns are as much directed at changing the attitudes among people as they are at influencing policy making. Such is the case, for example, with LGBT+ rights campaigns (lesbian-gay-bisexual-transsexual, plus spectrums of sexuality and gender). In many cases, activists and advocates actually put in practice a radical interpretation of democratic principles and demand that they be incorporated into the political system.

Perhaps advocacy networks' major efforts are aimed at shaping collective exercises of ethical imagination. Inwardly, network members build a sense of agreement in their plurality. Outwardly, they publicize messages, ideas, and information that together provide coherence and strength to their collective effort. The accumulated experiences of all those involved provide a fertile ground for a combination of resources and repertoires with which networks are able to position various issues on political agendas. This can take place at local, regional, national, and international levels. Through tactical uses of mainstream and alternative media, network members may achieve the increased awareness—visibilization—of topics and dissent regarding the status quo. In turn, this capacity to introduce issues in the public sphere opens up debates and exercises of collective imagination about what is possible. It also helps many people identify problems in aspects of their everyday lives they considered normal. In combining numerous activist groups, advocacy networks have thus an increased potential to include all stakeholders involved—or at least interested—in a certain issue. The various scales of their practice, from small localities to international policy arenas, also help translate specialized languages and local forms of knowledge for all those involved.

These processes are complex combinations of scales ranging from the micro or intersubjective to the macro or mass mediated. Some repercussions of campaigns, debates, projects, or demonstrations may end up being quite removed from activists' and advocates' expectations. But because advocacy networks imply constant negotiation between so many different actors, or at least deliberations among a few of the networks' groups that must consider the plurality of the networks themselves, there is more room for constant fine tuning and planning. Perhaps those campaigns that are better attuned to cultural translation and more prone to allowing multiple populations to appropriate them can reach farther. Some networks' campaigns or collective efforts have benefited from internal debates that have resulted in cultural sensitivity of network members, which has in turn led to policy changes or changes in the behavior of populations. What is at stake is the aim of changing expectations about what behaviors would be appropriate for state officials and the general population. Because such changes cannot occur overnight, the principle followed is that of a critical mass. What is meant here is not at what point people decide to mobilize (Oliver and Marwell 1988), but rather how a change in political clout requires that enough people back the changes sought in order for these to enter as potential aims in the political arena among state officials. In essence, it is a type of organizational behavior that seeks to alter the shared ethos through a garnering of support. But this then goes into the way political scientists have analyzed opposition to government policies, for example, regarding women's rights (Childs and Krook 2006; 2009).

For Douglas and Wildavsky, "moral opinions are prepared by the social institutions" (Douglas and Wildavsky 1982, 120). Both authors argue that it is rare and difficult for individuals to choose a moral stand based solely on individual rational grounds. In their perspective, justice, for example, needs to be balanced between the concept of agency and of community: "If, in the theory of justice, the so-called community is of a kind that never penetrates the minds of its members, if their shared experiences within it make no difference to their wants and contribute nothing to their self-definition or to their ideas of merit, then much is wrong with the theory" (Douglas and Wildavsky 1982, 126). This does not mean that community overrides all other considerations, but rather that there must be a sense of a shared belonging for a consideration of justice among all those involved to exist.

I consider the concept of community problematic, but I agree with Rapport when he says that community is "good to think with" (Amit and Rapport 2012). With cosmopolitan ideals, the sense of community expands to reach across borders. The challenge thus becomes to find formulas to engage with people who have contrasting histories and cultural contexts. The World Social Forum (WSF) is a good example of a place in which facilitators have tried not to guide topics or issues, but rather to provide an arena for people to come together and talk about their plans as well as to possibly learn from one another or come up with new ideas. The WSF is therefore an umbrella organization made up of interactions among numerous groups who seek to put social issues back on the international political agenda. Its slogan, "Another world is possible," is meant to be an optimistic message that will allow participants and others to work on achieving just that: another world. As a platform for the visibilization of the variety and complexity of progressive movements and campaigns, the WSF claims to lay the groundwork for what Alexander refers to as a "civil sphere"—"a world of values and institutions that generates the capacity for social criticism and democratic integration at the same time" (Alexander 2006, 4). In practice, however, what it does is connect people. It also makes room for the awareness of other campaigns, other projects, and other issues to be known outside the WSF's remit.

In doing so, the WSF combines the scales I was referring to earlier. On the macrolevel, it uses mainstream and alternative media to send out reports on its various meetings. On the mesolevel, the networks that conform to it share their work with the guidelines provided in its charter of principles. On the microlevel, activists and advocates are faced with a challenge of interpreting foreign ideas and establishing dialogues with potential allies. The cumulative effect of such interpersonal experiences reaches far and wide, both within advocacy networks and beyond. This chain of significations replicates those that have shaped cultural understandings throughout centuries. As Bloch points out, "Human history is the process of social interpenetration between individuals with the added twist that the representations involved, although in a state of transformation, can be the basis of further transformations of representations occurring in long chains of communication through generations or between contemporaries. Thus, information from long ago can either be reproduced or become the basis of yet further innovations and representations" (Bloch 2012, 177). Because the purposes of advocacy and activism entail an outright modification of shared norms and institutional policies, the awareness sought in supporters—about injustices in need of solution or about necessary changes—is of a kind that requires cognitive engagement.

The type of activist engagement referred to here is therefore not so much an individualist or holist approach, but rather a connectionist one, as Connolly describes in his interpretation of William James's work: "Connectionism overcomes the simple problem of induction while introducing a more profound one that haunts life periodically; it encourages us to infer from connected experience, while remaining alert to possible surprises that may overturn some of those inferences" (Connolly 2011, 35–36). Bloch explains that connectionism is a controversial neurological theory that purports that we take information simultaneously from different types of perception (Bloch 2012, 195). "The connectionists suggest that instead of knowledge being stored in linear sentential fashion it becomes reorganised in webs of networks connected in a multitude of ways" (Bloch 2012, 195). It would perhaps be more applicable here to refer to meshworks, as Ingold does, due to the relevance not only of the connections themselves but also of the medium where they take place, as well as their trajectories (Ingold 2011, 63–70).

The fact that religious organizations are very much directly involved with advocacy efforts, and sometimes discreetly so with activists, is proof of a combination of moral principles that feed the multilayered experience of those involved. Advocacy networks thus are able to combine a wide range of groups that allow for a certain type of exchange to take place: of hopes, dreams, and aspirations of potential futures. These may incite angry responses from some powerful actors to actions that are deemed harmful to the attainment of those dreams or to positive attitudes essential for actions that shape innovations in political practices at local, regional, or international levels. What they all point toward, though, is a process of becoming some type of collective conscience or moral compass through which activists hope to illuminate the path of wider assemblies.

In this chapter, I analyzed the negotiations of principles and aims within advocacy networks as "moral entanglements." During my fieldwork in both Spain and Brazil, I frequently noticed that the tensions between groups were dealt with in a practical fashion. Activists and advocates constantly sought to first establish good practices as ethical, to later identify the shared moral principles that allowed for such practices to occur. With this process, the moral compass that advocacy networks often seek to become is somewhat complexified as not a straightforward set of established principles but rather as a constant negotiation between vernacular and cosmopolitan principles. In seeking collective decisions, activists and advocates often engage in what I have termed "ethical liaisons," which help clarify the basic agreements of what good practices mean for all participants in a network. Because networks are not static organizations, they constantly deal with changing members or situations, thus making the result not a final aggregation but a type of dynamic moral stance. This in turn allows for a collective vision of the future, with which all participants can agree. In sum, this process allows for the final element toward civil becomings, which is the political project that I develop in the following chapter.

8

Democracy Reimagined

The [socioenvironmental] movement has a difficulty to establish a relationship with public institutions, like with an opportunity as this one [Pronatureza], they fight it . . . but also they have a capacity to make strong criticisms but real difficulties to make suggestions.

—Luiz, government coordinator of Pronatureza at the
Brazilian Environmental Ministry

ON NOVEMBER 16, 2004, I arrived alongside a small Instituto do Meio Ambiente da Amazônia (IMA) team at a retreat space outside of Manaus. It was an enclosed area that included an auditorium, several working rooms, a roofed terrace that doubled as an eating area, and several sleeping quarters, mostly large rooms with bunk beds. The area belonged to a religious order, which used it and rented it for different events. Over the next four days, we took part in an intensive workshop with technicians from numerous NGOs and local governments. The topic of discussion was a project I will call Pronatureza, which was started by IMA and was incorporated as government policy for the Amazon region by the federal government. It had been developed by a network of NGOs—led by IMA—for three and a half years before the government took it over a year before the meeting. Its purpose was to help small landowners comply with different laws and regulations regarding the protection and restoration of standing forest in combination with the system of rewards for environmental services. The key element of the project was that small landowners would receive money for protecting the forest on 80 percent of their land. Other aspects included special workshops to improve farming practices and the commercialization of produce. It also involved several local NGOs in different areas of the Amazon region to help with assessments and workshops for farmers. A few hundred landowning families were part of the pilot program, whose results and challenges we would be analyzing. Most sessions were discussions in an air-conditioned auditorium, where local teams reported on progress, challenges, and outlooks. One day all participants separated into several teams and each visited a few participating farms not too far from Manaus. At the end of that day, we all reconvened and discussed the results. Pronatureza was, in its design, closer to an advocacy network campaign, so this presented challenges for its proper functioning within the government. On another occasion, Nathan had shared his skepticism about the potential of the project: "I believe that Pronatureza was taken over by the government too soon . . . we could have worked on it a bit longer for it to work better."

This chapter is an exploration of the manner in which advocacy networks struggle to reimagine democratic ideals of collective forms of decision-making. In both the Amazon and Barcelona, the groups involved in networks sought improved forms of governance over shared or public issues. They did so through an openly or discreetly critical stance against state government policies. Both stances, nevertheless, implied an engagement with state institutions. In doing so, their members sought to enact their belonging to the recognized sphere of action known as organized civil society. They sought legitimacy in the eyes of states (Kamminga 2007). At the same time, state governments and interstate institutions have specific policies for dealing with civil society according to the prevalent ideological stance within each region. Those states that claim to uphold democratic principles and institutional mechanisms highlight the relevance of civil society organizations to democracy. Moreover, they have a series of policies in place to support them. This is in contrast to states that are skeptical about democracy; these tend to restrict the activities of independent groups and seek to control them. In recent years, several states like India, Egypt, and Russia have severely limited the scope of action especially of foreign NGOs (Rutzen 2015). In Brazil, I noticed a certain limitation on international NGOs, but which meant they only had to establish working groups in the country in order to establish rapport with authorities. That is why most major international NGOs have had large national chapters in the country for longer than in other countries. But all activists and advocates claim their roles as essential to the democratic life of the country where they are based. As mentioned, both field sites in this book, Brazil and Spain, went through long periods of authoritarian regimes in the twentieth century. The democratic forces that prevailed after the fall of the military regimes in both countries were adamant about stimulating an active civil society.

Pronatureza is an example of engagement between government and advocacy networks. As a project devised and developed by NGOs in the locations where it would take place, their firsthand knowledge of the areas involved, their inhabitants and ecosystems, were invaluable. Its first phases, even while coordinated by the government, were funded by international foundations. This is especially significant as the Amazon is a region where the Brazilian state is spread very thinly. Luiz was in charge of Pronatureza at the government Environment Ministry. Before joining the government, he used to work in IMA where he had been involved in Pronatureza's design and development. He thus had intimate knowledge of its mechanisms and of the aspirations that had shaped it. But for Marcio, IMA's technician who was in charge of the NGO's engagement with the project, Luiz lacked field know-how. After the workshop, he told me that he did not think Luiz could implement effectively what should be a project that engages directly with those who live in the forest. "He's a city kid . . . look at him. Did you see how he talked to the farmers?" In his early thirties, Marcio had long experience working in NGOs in the field. He easily switched from specialized technical language to banter and have fun with locals. He was visibly comfortable with farmers in very humble settings. His clothes were simple, yet sturdy: a black T-shirt and jeans. Luiz, however, was noticeably uncomfortable with the heat and dust of the farms. He was wearing a formal shirt with brown chinos, an outfit that seemed out of place as we walked through the different farms. He was much too formal when he talked to farmers, to which they often reacted with a somewhat bewildered expression. He appeared to be more comfortable among high-level bureaucrats than with technicians and farmers. After the last dinner of the workshop, he and I spoke for a while. "Once in government, Pronatureza increased its size and networks, as well as responsibilities," he told me, while also explaining the types of problems that had arisen among different ministries. "With the Agrarian Development Ministry we now have a problem on territories [land tenure], which is out of our control. But we do not intend to control these. We want to integrate what they are doing . . . but we have to deal with that." From what he explained, the complexity of Pronatureza made it easy for misunderstandings to travel quickly and cause friction among participants, especially NGOs and social movements: "The [socioenvironmental] movement has a difficulty to establish a relationship with public institutions, like with an opportunity as this one [Pronatureza], they fight it . . . but also they have a capacity to make strong criticisms but real difficulties to make suggestions."

All activists I met were often engaged in some way with or against state, regional, or local governments. The state, thus, was essential to their work. That for many groups the key demand is better governance of a wide variety of issues is proof that they seek to influence the way in which governments manage collective affairs. Governance is a term that has mutated from its original meaning of "action or meaning of governing" (Oxford Dictionary, 2001) to one that entails the collectivization of legitimate public decision-making for the common good (World Bank 2007). The way in which the term arrived at its current understanding was through descriptive reports of "bad governance," in which the report writers were evaluating the way in which decisions were made within organizations or governments. "Good governance" emerged as a response to such negative appraisals and became the ideal to follow. Foundations and international institutions promote it among their beneficiary organizations; national governments stimulate it at different levels of polity; corporations advertise the fact that "good governance is good business" (Governance 2007); and the groups identified as civil society or third sector have adopted it as their flagship concept to improve standards of management and lobbying. Eventually, the term evolved into its truncated form, "governance." It is now common to find the term in a wide range of academic disciplines as representing the ideal organizing principle.

For some critics, however, governance actually follows a pattern of neoliberal policies, inasmuch as it represents a privatization of decision-making that affects society at large. For these critics, some of the projects claiming to improve the quality of public decision-making actually neutralize strong political demands by passing on quotidian responsibilities to inexperienced groups. This follows the pattern of antipolitics and depoliticization that has been identified as an effect of increased involvement of NGOs in public policy making (Ferguson 1994; Fawcett et al. 2017). Some of my research subjects, though, hoped that it may also prove to be some sort of boomerang. By being handed the opportunity to determine what is good governance, people may be provided with the technical and policy know-how and experience in collective decision-making and dealing with state bureaucracies that will let them proceed pragmatically. Both attitudes toward governance, the critical or cautionary and the hopeful and optimistic, are similar to what has taken place regarding the concept of "empowerment" within development studies (Cheater 1999).

It seems ironic to think that independent citizens' groups called "nongovernmental organizations" actually reify the government as they plead for more state intervention in a series of issue areas. Within the advocacy networks I observed, however, they are not the only ones raising such demands. Social movements, unions, academics, and other organizations that are also part of the advocacy networks I describe would constantly refer critically to the state not because it existed or they opposed it altogether, but because they considered that it was not doing its work properly or fully. This engagement put in evidence a tension within some groups. A few activists and advocates I got to know were publicly extremely critical of the state, but their actions and, moreover, the actions of the organizations they were part of, worked in favor of the state rather than against it. An example of this phenomenon was a demand I witnessed from various groups within a network for increased policing and territorial control in the Brazilian Amazon. "We need better protection from the state against the violent loggers," said one activist during a meeting. In this and other events, the organizations I observed had a clear idea of what should be done to improve the state's services in the issues mentioned. Their campaigns seemed more focused on engaging the government to achieve what the organizations saw as necessary changes to policies than they were on the population that was affected by specific problems. Thus, I believe that advocacy networks actually form a sort of "shadow state" that complements the government's work and pushes it in directions that it would not have explored by itself.

An example of this is in advocacy networks' work on the definition and interpretation of norms and laws, which means that they are increasingly influential in the "the creation, institutionalization, and monitoring of norms" (Khagram, Riker, and Sikkink 2002, 12). Yet, by contesting existing norms and trying to influence their improvement and enforcement, an organization is actually legitimizing the contested institution. Nelson (2002, 149) provides an example of such a situation with the World Bank, which he argues has actually benefited from the protests against it by environmental advocacy networks. Instead of weakening its authority, such efforts have "in effect worked to strengthen its regulatory obligations and its power to regulate borrowing country governments" (Nelson 2002, 132). This actually helps a particular campaign or struggle while strengthening the authority of the state. The fact that strong norms already exist allows for campaigns and networks to be much more effective. According to Hawkins (2002, 49), the Chilean case of human rights mobilization throughout the Pinochet dictatorship proves this point. Ideas of democracy and human rights were already strong within Chile and in the international community, a fact that helped the campaign against the regime. The result was a thorough documentation of human rights violations, an international effort to impose sanctions on the Chilean government, and enough international attention to those contesting groups within the country to avoid their outright repression.

This means that in seeking their place within political systems, activists and advocates have formed networks to achieve united fronts that can have more influence than individual groups. An example of this is the labor movement, of which Kidder (2002, 290) argues that the "transnational nature of capital . . . demands a transnational response from labor" organizations. Unions fight for the states to enforce their rights and ensure their well-being against the private interests of their employers. Their situations are helped by cross-border fertilizations of ideas, campaigns, and support. It is interesting to remember that some of the first transnational advocacy efforts were workers' organizations in the form of the Socialist International (Nimtz 2002). In Latin America and Spain, unions are also famous for being easily co-opted. In both field sites, the politics of labor unions were closely related to state priorities. In Brazil, for example, the fact that the president during my fieldwork, Luiz Inácio "Lula" da Silva, belonged to the Workers' Party (Partido dos Trabalhadores, PT), minimized union contestation against the government, although not the discontent felt in some sectors of the labor movement. The unions I encountered in the Amazon were of smallholding local farmers. I talked to several union members, and the sense I got was that they appreciated belonging to networks where they learned to understand their individual plights in the context of larger dilemmas that the networks addressed. Activists often persuaded union leaders to join networks and helped them to navigate the use of sophisticated environmental discourses necessary for more fully engaging with projects and campaigns. I once joined Mariana— one of IMA's political operators—during her rounds to visit several farmers' unions scattered throughout the state of Pará. I noticed how a document farmers were supposed to write was actually typed by Mariana on the computer the network had donated to the union.

In Spain, the situation was different. The socialist government had been in power for a few years at all three levels of government in Barcelona (federal, regional, and local). Javier told me that most of the unions had an agreement with the government to avoid any conflicts and negotiate any problems on the side. However, CATAC, the independent union he led and which was strongly involved in the organizing committee of the Mediterranean Social Forum (Fòrum Social Mediterrani, FSMed), was extremely critical of such proximity. Indeed, it had called a strike in one of its branches against what it denounced as an injustice. Javier showed a similar challenging attitude in the FSMed Technical Secretariat, where he openly accused the other unions involved in the committee of seeking to "neutralize" any strong political message he tried to push through. This was particularly the case with the largest union in Spain, Comisiones Obreras, whose delegate in the FSMed was accused on several occasions of not even publicizing the FSMed within the union's membership (another committee member was a CCOO union affiliate and frequently complained he had received no information from the union's internal publications or newsletters). This example shows how some of the network members can perform quite differently either from their ideals or from the idealized image they have of themselves. This can help or hinder their position with respect to the government as well as with respect to their movement and stated collective aims.

It is common, therefore, for networks to become training grounds for local groups to be confronted with contrasting visions. What ends up prevailing is the overall aim that all network members share. Many cross-border advocacy networks claim to strive for a "global reality checking" (Ritchie 2002, 295), which entails a comparison of situations that helps set standards for debates and campaigns. For this reason, states sometimes consider networks and their members a threat. Some governments have clearly accused NGOs and their networks of being means for external intervention in their internal affairs. An example of this can be found in Indonesia, according to Riker (2002, 194), where the federal government considered NGOs to be "undermining the authority" of the state or "upsetting [its] political stability and security." Other state governments, like those of Russia and China, have applied strict controls on NGOs for fear of their political uses against the regimes. This type of measure may be a reaction against the campaigns themselves and the campaigners' efforts to make state governments comply with international norms.

Another strategy aimed at producing discomfort in authorities and that is frequently used in human rights issues is the "mobilization of shame." It is, according to Khagram, Riker, and Sikkink (2002, 16), a strategy exercised by advocacy networks to embarrass public authorities and private firms on the international stage either to force them to conform to norms or to show the inadequacy of the norms at hand. It is a play between the meaning of the ideas as reflected in the norms and the issues that have a direct impact on the population. This has been put in practice in many different issue areas, such as that of dams. In terms of meaning management, Khagram (2002) shows how certain campaigns helped change the understanding of dams from being considered as symbols of development and modernity to being perceived as controversial and highly problematic constructions (Khagram, Riker, and Sikkink 2002, 13). Part of the appeal that the campaign against the Narmada Valley dams in India achieved in the international community was the interconnections it showed between Dalit rights movements, human rights, and the environment (Khagram 2002, 208). It is significant that an alternative performative politics relies on shame to sway political decisions, to "convert sentiment, anger, and assertion against dominant institutions into effective and sustained political strategies" (Kothari 2002, 239).

In its Brazilian networks, IMA played a diplomatic role that helped tone down contestation from some groups to reach productive agreements with government officials and other powerful actors. That role came with some attitudes that were deemed unhelpful by some network members. Nathan, for example, had a clear vision of what IMA's role was at the time of my fieldwork in Brazil: "What we are in right now is the whole idea of leveling the playing field: giving social movements, grassroots movements fire and tools, the information in maps and projections that we give, to bargain on an equal basis." His use of "fire and tools" was of course reminiscent of a missionary compulsion to bring God's truth to communities. His was a common attitude among some of IMA's scientists. In practice, however, they participated in meetings where the results of their experiments and studies were discussed with communities, where those people's experiences and viewpoints were considered. The type of pressure that IMA is able to exert on the Brazilian state is double: mobilizing grassroots movements and organizations, and also mobilizing international foundations and aid agencies. By providing both types of networks with scientific data and analyses, IMA plays an active role in providing evidence that helps put pressure on the Brazilian state government to do something. In an interview at his house in Belém, Tony referred to how pressure by civil society organizations works on state officials: "Well, there is an ambiguity in Brazilian civil society, the Brazilian government, and their relations with internal and external influence . . . [which] is really not that different. . . . I remember the head of the MST once saying—the government was complaining about invasions—well, 'that is the only way you operate, it is not possible to negotiate realistically with you without taking action. You don't do it.' And there is a certain amount of the same thing in the international pressure the Brazilian government and people don't like but to which they respond; whereas internal pressure they tend to ignore. So basically they [government officials] complain about colonial attitudes, but their behavior is colonial."

IMA and the other groups with which it forms networks for projects or campaigns, therefore, see themselves as representatives of not only a Brazilian civil society but also, crucially, of a global one, to exert pressure on the Brazilian state. With their campaigns and collective efforts, advocacy networks try to change not only policies but also behaviors of the wider population. In doing so, they are seeking to shape the type of democracy they aspire to. Vernacular responses to democracy have yet to be better understood and are becoming of increasing interest among anthropologists (Paley 2008a). If government institutions work within a framework of microphysics of power (Foucault 1991, 26–27), then these groups seek to counter their influence through a similar exercise—but with the added value of seeking a legitimacy from outside the institutions. As we have seen, however, this does not necessarily mean they work outside of the state, but rather that they desire for populations to reclaim their stakes on the state. It is a true effort of appropriation. An elementary aspect of this process is that knowledge is produced in such a way as to challenge that which is used by state government institutions. And because "power and knowledge directly imply one another" (Foucault 1991,

27–28), then these groups are able to wield power. This is particularly obvious in the Brazilian Amazon, where IMA successfully produces knowledge that influences government policies. For Foucault, ceasing to describe the effects of power in negative terms was imperative to better understand the dynamics: "In fact, power produces; it produces reality; it produces domains of objects and rituals of truth" (Foucault 1991, 194). States established regimes of truth in order to centralize and manage reality within the confines of state polity. Such was the case of the judiciary, for example. Foucault sought to understand such processes in history. Through the systematization of discipline and its incorporation in the institutional design of governments, the bureaucratic modern state took shape.

What has taken place over the last couple hundred years, however, is a "de facto global governance" in several issues where advocacy networks have become very active (Khagram, Riker, and Sikkink 2002, 4). This fact has led many scholars to refer to an active "global civil society" (Taylor 2004, 2002). Others have developed theoretical approaches by which they try to make sense of the participation of NGOs, social movements, and advocacy networks in the international political arena (Kelly 2007). From the perspective of these approaches, such groups form part of what Gustavo Lins Ribeiro calls "alter-native transnational processes and agents" (Ribeiro 2009). In trying to establish a global moral compass, advocacy networks and their members have pushed for topics on which wide coalitions can be built. Some of these draw strength from religious organizations, such as campaigns to diminish the debts of poor countries (e.g., Jubilee 2000) (Busby 2007) or to reduce poverty (e.g., Make Poverty History) (Nash 2008). Others build on demands for rights. The institutional framework of international organizations such as the United Nations and its agencies is said to have contributed to a transnational governmentality (Ferguson and Gupta 2002). As we have already seen, for many critics, the notion of civil society is inherently linked to neoliberal policies, which means that their campaigns function as strategies to contain and control state populations (Baker-Cristales 2008, 351). Scholars who follow this viewpoint are indebted to Antonio Gramsci, who referred to the state as something that is formed by political society *and* civil society (Gramsci 1971). In transnational space, therefore, it is easy to understand a fear of a neoliberal governmentality.

In humanitarianism, for example, international legal arbitration functions with a series of strategies that appear to be more linked to the venture capital world (Dezalay and Garth 1996). This institutional bias toward corporate capitalism is, Rabinow explains, a form of "new symbolic imperialism": "*To invest in civic virtue is also to construct the state and to assure oneself of a position of legitimacy on the international market of savoir d'état* [state knowledges]" (Rabinow 2007, 49, emphasis in original). For state actors, many NGOs actually constitute a moral intervention that is less costly than wars. For this reason, humanitarian NGOs are known among state officials as "the most powerful pacific weapons of the new world order" (Rabinow 2007, 50). This perspective provides arguments for the very skeptical position regarding NGOs and advocacy networks. It appears to assume that a specific sense of morality can be transferred onto a social realm where it did not previously exist. The basic problem of this premise is that morality is processual and may be understood as occurring in two spheres: in practice and in discourse. In the former, morality occurs as "a kind of habitus or an unreflective and unreflexive disposition of everyday social life," which is "not thought out beforehand, nor it is noticed when it is performed. It is simply done. It is one's everyday embodied way of being in the world"; in the latter, morality occurs as "publicly articulated" ideas of what is good, appropriate, and expected (Zigon 2008, 17–18).

Instead of assuming the worst from advocacy networks or their member groups, some authors focus on what they see as their true potential. Arjun Appadurai turns Ferguson and Gupta's formulation around and speaks of "governmentality from below," or "countergovernmentality" (Appadurai 2002). This opens the door for a real risk among some scholars of assuming that advocacy networks are inherently good. An example was a study of "new social movements" whose authors titled their essay "Networks that give liberty" (*Redes que dan libertad*) (Riechmann and Fernández Buey 1994). Edelman has responded with caution to such optimism, pointing out that networks are simply organizational arrangements that may allow for many solutions to practical problems (Edelman 2005, 31–32). There are clearly criminal, militant, or terrorist networks that abuse networks' open structures to engage in "netwars" (Arquilla and Ronfeldt 2001). It is worth remembering that as a form of institutional architecture, advocacy networks mimic a transition in organizational arrangements that started among corporations and state governments. Through their change, these organizations sought to distribute tasks according to an area of specialization that puts in practice Weber's principle of bureaucracy as a rational form of governance (Goodsell 2005). Such a path owes much to Weber, for whom bureaucracy became "the organizational form of modernity" (Reed 2005, 117), which required an orderly distribution of areas of expertise.

Wolf identifies four modalities of power: "who can talk, in what order, through which discursive procedures, and about what topics" (Wolf 1999, 55). If power conforms to these modalities, then by making their demands and complaints visible in mainstream media, advocacy networks are exercising a very real power. When advocacy networks widen conversations about shared issues, the populations involved may more easily find common moral principles (Mulgan 2006, 235). Just as democracy has become a legitimating principle of government through which even "dictators routinely excuse themselves on the grounds that they are simply stabilizing matters to lay the ground for democracy's return" (Dryzek 1999), so too does the idea of governance increasingly play a legitimating role in public affairs. Perhaps Warkentin's definition provides a valuable insight: "*global civil society is a socially constructed and transnationally defined network of relationships that provides ideologically variable channels of opportunity for political involvement*" (Warkentin 2001, 174, emphasis in original). In this view, global civil society is not an actual assemblage but rather a sphere of connectivity that allows for opportunities to engage with the political. It is a sort of transnational public space (Avritzer 2002, 6).

Of the four modalities of power that Wolf identified (Wolf 1999, 5), of which two are related to the individual and one to structural domains, there is one that he defines as "power that controls the contexts in which people exhibit their capabilities and interact with others" (Wolf 1999, 5). He calls it "tactical, or organizational power." The relationships that are encouraged through activism and advocacy seek to reconfigure the social. Through discourses and actions, activists and advocates hope to influence not only state government officials but also the wider population of the polity in question. Both networks analyzed in this volume sought to serve as initiators of change. The ones in the Brazilian Amazon did so through the mediation between conflicting interests regarding the forest and its territories. The ones converging in the Mediterranean Social Forum (FSMed) sought to provide spaces for interaction for groups living under totalitarian regimes in order to promote their struggles. In both cases, what was sought was a sphere of interaction, where dialogues about public issues could take place. All those involved thus sought to reimagine democracy and how it could be put in practice in different contexts.

Democracy has come to be understood as a multilayered and plural set of practices and institutions that together seek to involve wider populations in collective decision-making processes about issues that will affect them. The varieties of institutional arrangements and legal frameworks that are put in practice in its name reveal the constant efforts to balance functionality and legitimacy (Hendriks 2010). As a symbol, it has been used by activists and advocates to justify their involvement in numerous issues as well as fight totalitarian tendencies in governments. It has also been used to try to motivate those that NGOs and activist groups try to help. In this sense, the activists and advocates I met in Brazil and Barcelona constantly reminded farmers, indigenous communities, rubber tappers, and other activists about their right to be heard, in their complaints and suggestions, in order to modify government policies or projects. But it was not merely a case of empowering stakeholders. I attended several meetings in Brazil in which IMA's workers sought to make forest dwellers understand that they were part of a more complex ecosystem, which also limited what they could ask for or expect. I consider this a cornerstone of how activists and advocates reimagined democracy: to constantly refresh the view on the collective to not lose sight of its significance for individual claims or expectations.

Pronatureza is a good example of a collaborative effort through advocacy networks in which complex interactions seek to solve existing problems and reach ideal goals. Luiz explained to me that although IMA designed Pronatureza as a project that could be scaled up, its origins were to be found in social movements. Some activists had learned about the potential of payments for environmental services (Pagiola, Bishop, and Landell-Mills 2012) and started developing a series of concepts to imagine an ambitious plan to reconfigure small-scale farming in the region. The four-day meeting I attended was evaluating the first year of trials of Pronatureza, which had included hundreds of small-scale farms in each of the eight zones of the Brazilian Amazon. There were teams for each of the zones, who had carried out research about the main needs and characteristics of the farms in their areas. One of the tasks that Luiz and other technicians asked of farmers during our visits was to draw two maps of their properties: one of how they were on the day of the visit, and another on how they saw them ten years in the future. Ideally, the farmer would understand the changes that need to take place in order not only to comply with regulations but also to improve the productivity of the family's plot. For some technicians, allowing farmers to do their own maps was disingenuous, as they did not fully grasp the ideal distribution of forest and different crops in their territories. "The [farmer] family can do the current map, but has difficulties thinking about a diagnostic . . . a technician should help them imagine," insisted one in the evaluation meeting afterward. After a few other similar complaints, Luiz intervened: "It is true that the producer is not yet understanding the aim of the program [Pronatureza]. One of the reasons is that we have changed our personnel too often. We need not only for them to change things, but to understand why those changes are necessary. . . . I think we are not yet where we want to be." The challenge Pronatureza faced was to link international efforts to offset emissions by supporting the benefit to the global environment that some ecosystems—like the Amazon forest—provide, with local small landowners who often cut the forest to make a profit.

The ambition of Pronatureza was to solve numerous problems all with one single project. There were numerous circumstances that those in charge had not foreseen. "Many farmers join the program to get payments for protecting the forest, but the first thing they may get from the government is a fine because they do not respect the percentage of standing forest that should be there in their land," said one American scientist who had been invited to give a talk at the workshop. "We are learning as we go along," Luiz insisted whenever something like this came up. What lies behind this project, however, is the aim to help local forest dwellers understand their role as fundamental for the Amazon region as a whole. The challenge was therefore to balance between the self-interest of farmers and the interest of the country in protecting the region's ecosystems. This mirrors Chris Hann's opinion that "the fundamental tension is that between particular and universal interests, between the selfish goals of individual actors and the need for some basic collective solidarity in a moral community" (Hann 1996, 3–4).

In Barcelona, efforts by the Technical Secretariat to make the FSMed event an ideal democratic space to inspire and motivate action in the Mediterranean seemed overshadowed by Catalan nationalism. Spain's symbolic democratic credentials after going through a dictatorship (Edles 2010) served as an ideal for other countries living under dictatorship. It was therefore paradoxical that a transnational arena that was thought of as a safe space for activists from around the Mediterranean would be hijacked by a nationalist agenda. The constant frictions that the FSMed process went through were directly related to the centralization of responsibility, resources, and decisions by groups based in Catalonia. For all their discourses on inclusion, plurality, and the value of collective decision-making, activists and advocates in Barcelona ended up excluding many groups and visions from its proceedings. Nevertheless, when the FSMed event took place, those activists who were there took the opportunity to network: to meet their peers from other countries, establish contacts, and seek to collaborate. It was therefore also an example of activists making the most of a nonideal situation. Despite the fact that the FSMed event failed to convince its proponents to continue with a similar format as a continuous process, two editions of the World Social Forum were later held in Tunisia (2013, 2015). This was due to the fact that Tunisia is widely considered as the only successful democratization deriving from the uprisings known as the Arab Spring (Masri 2017). So even if the FSMed process itself dissolved, the individuals who took part in it learned a few lessons. When one network disappears, what follows is not a reluctance to participate in networks, or no-networks, but rather new networks where participants can put in practice what they have learned.

In both Brazil and Barcelona, activists and advocates faced difficulties in achieving democratic ideals but always acknowledged the fact that process matters more than labels. There was no end point, no summit to reach, but rather elusive goals: an improved socioenvironmental governance in the Brazilian Amazon, and functioning networks between progressive activists and advocates in the Mediterranean. In seeking to transform practices, those involved engaged in alternative performative politics. They thus dared to experiment with different projects or styles of association to seek a renewal in political practice. In many cases, old established forms of association, like corporatism or clientelism, kept lurking around. But that is one advantage of the network as a political form: if enough members can call out creeping unwanted practices, then there is potential for these to be stopped before they overtake the whole process. Or, perhaps as happened with the FSMed, the network itself implodes and its remains form new networks.

In this chapter, I explored the way in which democracy serves as a guiding symbol to shape the alternative performative politics of advocacy networks, and as something to be renewed by activists and advocates. Both cases that I have analyzed in this book faced difficulties in establishing working processes that incorporated enough ideals that activists and advocates envisioned, like inclusion, justice, and others. In great part, the frictions that advocacy networks faced reflect the agonistic character of the political (Mouffe 2005). It was problems within the networks themselves that held them back from going further in what their proponents had envisioned. But the complications faced within the networks actually help those involved to prepare for what occurred when the network engaged with powerful actors outside. Because advocacy networks are transnational, a decisive element of their attempts at renewing democracy is the constant tension between vernacular and cosmopolitan values. Each set of principles is usually not merely a collection of abstract ideas, but rather forms part of an arrangement of materials, practices, and relations through which they are acted out and reproduced. In great part, the challenge in bridging such local and transnational arrangements often passes through diplomatic efforts to find terms that may accommodate both. If this does not work, and one is blatantly selected over another, then activists and advocates need to persuade all those involved that such selection does not betray their beliefs or aims. These practices make activists and advocates master persuaders, as they need to become so to address the networks from within, as well as with those such networks seeks to address. All of this is usually carried out under the banner of the renewal of democracy, as a symbolic ideal and goal. Such constant enactment of a reimagined democracy in practice, whatever form it may take, also plays a crucial role in building the legitimacy of the network and its member organizations as respectable members of organized civil society. It is thus a fundamental part of their civil becoming.

Conclusion

Nonlinear Political Developments

The forums [the World Social Forum and its many iterations]
are nothing more than instruments, which help spread [our]
ideas and political positionings and help to create networks; but
because of their own plurality, and because it is not possible to
agree on joint fights or mobilization, or concrete actions, they
continue to be simply a broadcaster of social movements, and
that is important in itself.

> —Javier, labor union leader and member of the FSMed
> Technical Secretariat

CIVIL BECOMINGS REFERS to an ongoing process through which activists
and advocates continuously perform what they consider to be their roles as
civil society actors. In doing so, they are not only pursuing the goals set by
the organizations and networks in which they participate but also, crucially,
advocating for their own role as essential to the democratic project. With this
process, they are aiding in the imaginary construction of a political community. This means that through efforts that do not form part of official state bureaucracies, the idea of the political communities that provide the legitimacy
for such political institutions is continuously shaped. The role of advocacy
networks is particularly inventive because of their entangled agency, that is,
their capacity to decide and act in a different manner from that of their member groups, and because of their constant exercise in mediating vernacular
and cosmopolitan values. These characteristics by no means indicate that advocacy networks solve the problems they address or succeed where single
groups struggle. But their work cannot be simply evaluated through a binary
of success or failure. Many networks, like the ones analyzed in this volume,

fizzle out, disappear. Those who formed them nevertheless feel compelled to form new networks, to apply the lessons learned in the previous iterations. It is therefore an ongoing process, a becoming.

Several scholars of social movements have long argued that activism is a communicational affair (Melucci 2003). In some instances, as the case in Brazil demonstrates, the key contribution that citizen groups make to public life is in the form of knowledge production and its dissemination. In others, as the Mediterranean Social Forum (Fòrum Social Mediterrani, FSMed) illustrates, it is the interactions, with all their tensions, friction, and resolutions, that become building blocks for other joint actions for public affairs. Melucci emphasizes the concept of "collective action" as an experiential affair through which numerous individuals reach a unified voice (Melucci 2003, 20). Drawing on my case studies, however, I argue that what is built is not, by a long stretch, any type of homogenized conception of a collective or a "we." Indeed, heterogeneity is a characteristic among activists and advocates specifically as well as in a wider sense (e.g., in the political community). In practice, activists and advocates, be they radicals or reformers, understand the differences between them and decide to collaborate even if they do not share some principles. Each activist group or NGO may develop its own cohesion, sometimes with similar motivations among its members, and other times, through specific agendas and aims.

Advocacy networks, it is argued here, function with an awareness of their diversity and an understanding that they necessarily must work together despite their differences. Furthermore, such plurality actually works in their favor. Numerous analyses of activism and advocacy have focused on a search for identity or homogeneity. This in turn has led to academic studies that portray activist and advocate groups as actors engaged in power struggles—through competition and negotiation for influence—with established institutional groups (e.g., government offices, international agencies, or political parties). The analysis presented here, however, argues that in the complexity of their collaborative practices—such as campaigns and political maneuvers—networked activists and advocates perform a nonlinear style of political engagements. By this, I mean that they do not seek to comply with conventional procedures to gain influence or dictate a political agenda but are rather disruptive and provocative by switching between types of questions asked: about what a problem is, and about its potential solution. This process also crucially involves a reflection of the situatedness of such a "problem," be it in a community, city, forest, and such. In doing so, the style of political engagement shakes up understandings of the types of public debates, identification of collective aspirations, common moral frameworks, and decision-making for governmental policy making. Perhaps my decision to group together two camps in the FSMed Technical Secretariat—of radicals and reformers—was not overly helpful for understanding the wide array of organizations and their individual influence on how the FSMed meeting finally played out. My analytic choice in doing so, however, responded to the ongoing conflicts and unspoken agreements that I witnessed. Although these sometimes took the form of sympathies and antipathies between individuals, they were also informed by larger issues. One example was the distrust among many of the participants of a Catalan influence in the FSMed; another was the apparent use by some Technical Secretariat members of the liaison with officials from the Spanish and Catalan governments for their own benefit. These fears were driving factors in the erosion of the enthusiasm needed to ensure further editions of the FSMed. By focusing on networks as the key arrangement of the activist practices presented here, I do not seek to romanticize their possibilities or capacities. Networks are not horizontal, although they may be inspired by horizontality as a model to achieve their aims. Likewise, they are not free from stark inequalities because these are organically formed, if not by material differences, by accumulated experience and knowledge. Such inequalities, nevertheless, do not stop the arrangement from signifying something new among those involved. For those involved, I learned from many conversations, networked collaboration meant an unprecedented level of possibility for influence.

The concept of nonlinearity is of crucial importance. In his introduction to *A Thousand Years of Nonlinear History*, Manuel DeLanda argues that chronological history has been written from a particular philosophical perspective that entails a linear understanding of events that accrue toward some sort of improvement (DeLanda 2011, 13). He argues for the need to engage in nonlinear descriptions that would produce different interpretations. By nonlinear dynamics, he refers to those "in which there are strong mutual interactions (or feedback) between components" (DeLanda 2011, 14). In his attempt to tell a tale of nonlinear history, he looks into the evolution of cities as material entities in which it is possible to refer to their dwellers as a sort of community. He refers to cities forming a meshwork, rather than a network, because it is not only a series of connections but rather an "interlocking system of complementary economic functions" (2011, 39). Tim Ingold has argued that the concept of "meshwork" is more effective than that of "network" to explain the entanglements with which life exists because its emphasis is not so much on the relations between elements/actors/objects but rather on the flows, mixtures, and mutations that these necessarily go through in any given contact or relation (Ingold 2008, 11). Meshwork therefore entails the feedback systems to which DeLanda already inferred. DeLanda argues that previous linear-causality analyses were the inevitable result of thinkers fitting the reality they observed into their theories. A nonlinear perspective thus provides analyses a less constraining framework in order to pay attention to the various paths of practices and events. In both field sites, the complexity of collaborative arrangements and their mixed results proved challenging to some participants who were used to less challenging environments. Such learning processes in plural networks, I argue, are crucial for renewed forms of legitimation or civil becomings.

As I waited once for Javier outside of his office in Barcelona's city center, I overheard several people approaching him with various cases that needed attention: one case was of a worker in a conflict over a workplace accident; another was of a local union asking for advice about a change in a key human resource regulation that the company sought to push through. After almost an hour, he finally came out and suggested we get a coffee in the bar downstairs. Once downstairs, he complained to me about the bureaucracy he had to deal with continually and praised the FSMed for helping him see the bigger picture even when the daily dealings seemed so mundane. Even though the FSMed encouraged him, he nonetheless had certain regrets about not being able to unify all participants into a type of joint action. He told me that "the forums [the World Social Forum and its many iterations] are nothing more than instruments, which help spread [our] ideas and political positionings and help to create networks; but because of their own plurality, and because it is not possible to agree on joint fights or mobilization, or concrete actions, they continue to be simply a broadcaster of social movements, and that is important in itself." Javier seemed simultaneously appreciative of the diversity within the FSMed and unhappy about its lack of unity. It was as if he reluctantly joined a struggle he did not consider to be a potential solution to a key problem (in his view, "capitalism") but nevertheless attempted to engage in wholeheartedly.

After spending several days in a meeting in a hotel in Alter do Chão, Brazil, several Instituto do Meio Ambiente da Amazônia (IMA) workers and I went to relax at the local river beach. It was very pleasant to end the day in the shallow waters of the riverbed, where families and friends were scattered in conversation or games, or were simply drinking. Some people had chairs halfway into the water so they could refresh their feet while they were seated. An ice cream seller pushed a floating cart among potential clients dispersed in the waist-deep water. While on the beach, I asked Adrienne, who had needed to defend herself after presenting a provocative proposal to activists, farmers, and environmentalists, "What do you aspire to with your work?" My question was itself deliberatively provocative. She smiled as she told me quite frankly: "You know, I just want to live a decent life with my partner, be able to buy a house, and enjoy my days. It is not easy in our field of NGOs or activism, but I think it is possible. Professionally, I know my work makes a difference, we do good stuff, but I do not want to sacrifice my quality of life for a cause. I am certain, though, that both are compatible." We had a couple of beers on the beach before going back to Santarém, where a few of them would take a flight back to Belém and I would stay. Every trip I took in the Brazilian Amazon reminded me of its vastness. The scale of the Amazon forest and its problems often made the region's challenges appear unsurmountable. But with their systematic work of tackling one issue at a time, through research, consultations, and meetings, IMA's personnel sought to help manage a change in practices that would aid in protecting as much standing forest as possible.

In both field sites, people involved in organizing meetings, setting up collaborative arrangements among diverse citizen-led groups, seeking to influence government policy making, and calling for attention to situations they defined as problematic, built complex arguments that relied not only on knowledge and language but also on images and performances. Each set of networked activists needed to constantly situate their efforts in their particular context. Some did it out of their own desire to have a positive impact on their region. This was the case with the scientists who decided to carry out extensive research in the Brazilian Amazon after they discovered that much of the original research that led to policies to protect it was carried out in Costa Rica. IMA's scientists' findings from research done in the Amazon itself would be more efficacious for protecting its forest. Others situated their efforts in a context that was related to external events or forces. Such was the case with the FSMed's Technical Secretariat, which needed to defend itself against accusations by foreign activists that they were Catalanizing the FSMed. The resulting process of negotiation was laden with a multilayered moral landscape in which activists struggled to reimagine democracy.

The present volume has sought to build an argument that helps readers to better understand the complex manner in which political engagements have changed because of what are known as advocacy networks. These, it has been argued here, have innovated in numerous fields of human endeavors by combining negotiations with government institutions and a constant yet ever-changing effort to remain as relevant actors in political arenas. Their efforts, however, cannot be considered in a linear, causal way. As easily as they gain influence one day, they can lose it the next. Understanding this process may help us better approach politics as a changing field of relations and influences that is more fluid and nonlinear than has been previously understood. David Chandler argues for a nonlinear way of understanding recent changes in political behavior around the world (2014). He sees the success of such a method as being reliant on an increasing number of subjects' stronger sense of agency. Historically, these subjects would not have been considered as deserving of a say in political decisions about their communities. He even says that "in a nonlinear world, the public must be understood as self-constituting through everyday decision-making and interaction" (Chandler 2014, 43).

This anthropological analysis examines the way in which activists and advocates develop symbolic tools they hope will help them change a wide range of situations, particularly regarding government policy making. Collaborations across national borders, bureaucratic orders, and other cultural boundaries have meant not a homogeneous appreciation of perceived problems and potential solutions but rather a multilayered plurality of sometimes contrasting and even contradictory aspirations and goals. In such circumstances, what occurs among those groups involved cannot be understood as an uncritical pursuit of Western Enlightenment values. Or rather, it cannot be considered as a Habermasian style of communication following a normative path of rationalization. It is rather a truly chaotic multilayering of various conflicting views and opinions where participating organizations nevertheless find a common ground on which they can collaborate. To view such efforts only through a lens of "success" according to a supposed common aim would be to remain blind to the reverberations of their actions. When such interests come into view, what prevails is an effort to fabricate stories of success, or what Mosse termed, in a study of development-oriented networks in India, the "social construction of success" (Mosse 2005, x). What Mosse described was a bureaucratic structure through which NGO-led projects were required to fulfil set criteria in order to justify their own participation in networks. This generated a series of practices that left many workers frustrated as their local efforts were subsumed by the wider goals of the network. It was part of the trend of ritualized verification that has multiplied audits in public and private organizations (Power 1997). This played more of a role in Brazil than in Barcelona, but it was present in both places.

Apart from any type of results, however, collaboration itself left wounds and joys, resentments and endearments, in both field sites. The conflicts, frictions, synergies, and outcomes of some of the efforts portrayed here perhaps helped individual activists and advocates—as well as the groups they formed part of—better understand collaborative efforts. Or perhaps they helped them clarify philias and phobias, aims and no-go areas, or aspirations and red lines. One fact is that most of those involved in the networks in turn went on to form new networks. So whatever lessons or experiences were learned, they were put to use shortly afterward. This is one of the feedbacks that occur in advocacy networks. Other feedbacks are either more individual or more collective. An interesting one relates to activism as a life decision. In their analyses of radical left activism, Krøijer and Maeckelbergh focused on conceptions of time and what those conceptions meant for activists' practices. Both, however, referred only laterally to the fact that many of their subjects of study were young individuals who dedicated their energies to their radical aims. Both appeared to suggest that the type of activism they focused on was only a life stage for most of those taking part. There were very few older activists among their interviewees, and those who stayed into their later years seemed to be linked to other forms of organization as well. Both in the FSMed and in IMA's networks, there was a wide variety of ages; however, crucially, some belonging to an older generation had transited through more radical groups in their youth. Although I did not do a systematic inquiry into their experiences, I can infer that for some of those who used to participate in radical and even violent groups in their youth, it was perhaps natural to move on to a more mellowed yet still progressive organization later on in life. This was definitely the case with Víctor, who sought to stay active in environmental organizations as he grew older.

The aim of this volume, however, is not to scrutinize what participating in advocacy networks meant for those who did so, but to reflect on wider implications for the political practices taking place in and through their midst. Such practices need to be situated within a longer history of aspirations for democracy. In seeking to shape collective responses to shared problems, especially by influencing various governments' policies, activists and advocates put in practice what they understood to be their rights. In his ambitious volume on the history of democracy, Keane (2009) provides a *longue durée* perspective on the contentious topic of "democracy." Careful to avoid any cultural simplifications, Keane demystifies some of the assumptions about democracy, such as its supposed birth in Greece. With care and rigor, Keane examines a long list of cases and examples that have contributed to some of the still multiple and sometimes quite polarized understandings that exist about democracy today. Throughout his volume, Keane reveals the fragility of democracy as a political arrangement. In doing so, he lays bare a nonlinear interpretation of its history. As he puts it in the introduction, "The exceptional thing about the type of government called democracy is that it demanded people see that nothing which is human is carved in stone, that everything is built on the shifting sands of time and place, and that therefore they would be wise to build and maintain ways of living together as equals, openly and flexibly" (Keane 2009, xii). Anthropologies of democracy have focused on the practices through which people in specific contexts seek to bring about their own ideas and ideals on the matter. What Keane's global and historical perspective contributes, however, is an understanding of institutions as fluid, as under constant renewal.

With their activism and advocacy, those involved in the networks portrayed here, as in similar ones around the globe, are renewing the way in which public decision-making works. By organizing groups and participating in networks, individuals perform their ideals for a democratic society. Their civil becomings are ongoing processes of deciding what is important for the broader social (and biosocial) collective they form part of. Over the last few years, we have witnessed a series of changes in national political landscapes in which traditional political parties have seen their support drop, sometimes dramatically. Most attention has gone to fringe groups that coalesce in new political parties; thus, there has not been enough reflection on the reshaping of the political society that is spearheaded by organized groups of various sorts. We should not forget that just as the state has a relatively short history as a bureaucratic institutional arrangement for dealing with collective affairs, so too do political parties have a similar timeline as collaborative efforts to try and manage ideas and ideals of what the collective should look like.

Perhaps activist and advocacy networks are functioning now as political parties did when they started to form: as emergent arrangements among individuals who seek to identify common interests and ideals despite the contradictions, frictions, and misunderstandings that take place within their collectives. The fragility of such a process, however, also means there is no guarantee that their becomings will remain civil.

Works Cited

Abélès, Marc. 1988. "Review of 'Modern Political Ritual.'" *Current Anthropology* 29 (3): 391–404.

Abramson, David M. 1999. "A Critical Look at NGOs and Civil Society as Means to an End in Uzbekistan." *Human Organization* 58 (3): 240–50.

Acosta, Raúl. 2007. "Managing Dissent: Advocacy Networks in the Brazilian Amazon and the Mediterranean." PhD diss., University of Oxford.

———. 2009. *NGO and Social Movement Networking in the World Social Forum: An Anthropological Approach.* Saarbrücken, Germany: VDM.

———. 2013. "Capacity Infrastructure in Brazil: Legacies of Participation in Christian Base Communities." In *Faith in Civil Society: Religious Actors as Drivers of Change,* edited by Heidi Mosknes and Mia Melin, 134–43. Uppsala, Sweden: Uppsala Centre for Sustainable Development, Uppsala University.

Agrawal, Arun. 2005. *Environmentality: Technologies of Government and the Making of Subjects.* Durham, NC: Duke University Press.

Agustí, David. 2002. *Historia breve de Cataluña.* Madrid: Sílex.

Alabart, Anna. 1981. "Els barris de Barcelona i el moviment associatiu veïnal." PhD thesis, Universitat de Barcelona.

Alexander, Jeffrey. 2006. *The Civil Sphere.* Oxford: Oxford University Press.

Allegretti, Mary Helena. 2002. "A construção social de políticas ambientais: Chico Mendes e o movimento dos seringueiros." PhD thesis, Universidade de Brasília–DF.

Alonso, Angela, and Raymond Clémençon. 2010. Review of "Environmentalism Brazil: Between Domestic Identity and Response to International Challenges." *Journal of Environment and Development* 19 (3): 247–51.

Alvarez, Sonia, Jeffrey W. Rubin, Millie Thayer, Gianpaolo Baiocchi, and Agustin Lao-Montes. 2017. *Beyond Civil Society: Activism, Participation, and Protest in Latin America.* Durham, NC: Duke University Press.

Alves, Antonio. 2004. *Artigos em geral: Arqueologia do recente, livro três.* Rio Branco: Grafica de Brasilia.

Amit, Vered, and Nigel Rapport. 2012. *Community, Cosmopolitanism, and the Problem of Human Commonality.* London: Pluto.

Anderson, Benedict. 2006. *Imagined Communities: Reflections on the Origin and Spread of Nationalism.* Rev. ed. London: Verso.

Anderson, Robin L. 1999. *Colonization as Exploitation in the Amazon Rain Forest, 1758–1911*. Gainesville: University Press of Florida.

Andrade, Paes de, Oswaldo Lima Filho, Edmundo Galdino, Paulo Delgado, Aldo Arantes, Ademir Andrade, and Augusto Carvalho. 1989. *Homenagem Póstuma ao Sindicalista Chico Mendes: Discursos pronunciados pelos deputados*. Brasilia: Câmara dos Deputados.

Andreu, Marc. 2004a. "Fòrum de letras." *El Periódico*, Barcelona.

———. 2004b. "Otra lectura de la 'marca BCN.'" *El Periódico*, Barcelona.

———. 2008. "Moviments socials i crítica al 'model Barcelona': De l'esperança democràtica de 1979 al miratge olímpic de 1992 i la impostura cultural del 2004." *Scripta Nova* 12 (270): 119.

Appadurai, Arjun. 2002. "Deep Democracy: Urban Governmentality and the Horizon of Politics." *Public Culture* 14 (1): 21–47.

Armstrong, Kenneth A. 2002. "Rediscovering Civil Society: The European Union and the White Paper on Governance." *European Law Journal* 8 (1): 102–32.

Arquilla, John, and David Ronfeldt. 2001. *Networks and Netwars: The Future of Terror, Crime, and Militancy*. Santa Monica, CA: RAND.

Atwal, Maya. 2009. "Evaluating Nashi's Sustainability: Autonomy, Agency, and Activism." *Europe-Asia Studies* 61 (5): 743–58.

Avritzer, Leonardo. 2002. *Democracy and the Public Space in Latin America*. Princeton, NJ: Princeton University Press.

Babels. 2013. Accessed June 11. www.babels.org.

Bailey, Frederik G. 1969. *Stratagems and Spoils: A Social Anthropology of Politics*. Oxford: Basil Blackwell.

———. 1998. *The Need for Enemies: A Bestiary of Political Forms*. Ithaca, NY: Cornell University Press.

Baker-Cristales, Beth. 2008. "Magical Pursuits: Legitimacy and Representation in a Transnational Political Field." *American Anthropologist* 110 (3): 349–59.

Balée, William. 1989. "The Culture of Amazonian Forests." *Advances in Economic Botany* 7:1–21.

———. 2006. "The Research Program of Historical Ecology." *Annual Review of Anthropology* 35:75–98.

Balée, William, and Clark L. Erickson. 2006. *Time and Complexity in Historical Ecology: Studies in the Neotropical Lowlands*. New York: Columbia University Press.

Barbosa, Luiz C. 2000. *The Brazilian Amazon Rainforest: Global Ecopolitics, Development, and Democracy*. Lanham, MD: University Press of America.

Barnes, John A. 1954. "Class and Committees in a Norwegian Island Parish." *Human Relations*, no. 7, 39–58.

———. 1969. "Networks and Political Process." In *Local-Level Politics: Social and Cultural Perspectives*, edited by Marc J. Swartz. London: University of London Press.

Barrena, Begoña. 2004. "El gobierno catalán se vuelca en el estreno de Orwell en el teatro Romea." *El País*, Barcelona.

Bauman, Richard, and Charles L. Briggs. 1990. "Poetics and Performance as Critical Perspectives on Language and Social Life." *Annual Review of Anthropology* 19:59–88.

Beck, Erin. 2017. *How Development Projects Persist: Everyday Negotiations with Gua-temalan NGOs*. Durham, NC: Duke University Press.

Bellin, Eva. 2012. "Reconsidering the Robustness of Authoritarianism in the Middle East: Lessons from the Arab Spring." *Comparative Politics* 44 (2): 127–49.

Bensch, Stephen P. 1995. *Barcelona and Its Rulers, 1096–1291*. Cambridge: Cambridge University Press.

Bermeo, Nancy. 2000. "Civil Society after Democracy: Some Conclusions." In *Civil Society before Democracy: Lessons from Nineteenth-Century Europe*, edited by Nancy Bermeo and Philip Nord. Lanham, MD: Rowman and Littlefield.

Bevir, Mark. 1999. "Foucault and Critique: Deploying Agency against Autonomy." *Political Theory* 27 (1): 65–84.

Bijker, Wiebe E., Roland Bal, and Ruud Hendricks. 2009. *The Paradox of Scientific Authority: The Role of Scientific Advice in Democracies*. Cambridge, MA: MIT Press.

Bird, Kate, and David R. Hughes. 1997. "Ethical Consumerism: The Case of 'Fairly-Traded' Coffee." *Business Ethics: A European Review* 6 (3): 159–67. https://doi.org/10.1111/1467-8608.00063.

Blakeley, Georgina. 2005. "Local Governance and Local Democracy: The Barcelona Model." *Local Government Studies* 31 (2): 149–65.

Blanchar, Clara. 2004. "La jaima acogerá 15 exposiciones de pequeño formato ideadas por ONG." *El País*, Barcelona.

Bloch, Maurice. 2012. *Anthropology and the Cognitive Challenge*. Cambridge: Cambridge University Press.

Blockmans, Wim. 2008. *The Medieval Origins of Constitutional Representation*. Europeaeum. Oxford: University of Oxford.

Boéri, Julie. 2008. "A Narrative Account of the Babels vs. Naumann Controversy: Competing Perspectives on Activism in Conference Interpreting." *The Translator* 14 (1): 21–50.

———. 2009. "Babels, the Social Forum and the Conference: Interpreting Community Overlapping and Competing Narratives on Activism and Interpreting in the Era of Globalisation." PhD thesis, University of Manchester.

Boggs, James P. 2004. "The Culture Concept as Theory, in Context." *Current Anthropology* 45 (2): 187–209.

Boli, John, and George M. Thomas. 1999. *Constructing World Culture: International Nongovernmental Organizations since 1875*. Stanford, CA: Stanford University Press.

Borja, Jordi. 1995. *Barcelona: Un modelo de transformación urbana, 1980–1995*. Quito: Programa de Gestión Urbana.

———. 2003. *La ciudad conquistada*. Madrid: Alianza.

Borja, Jordi, and Zaida Muxí, eds. 2004. *Urbanismo en el siglo XXI: Bilbao, Madrid, Valencia, Barcelona; Una visión crítica*. Barcelona: UPC ETSAB.

Bornstein, Erica, and Aradhana Sharma. 2016. "The Righteous and the Rightful: The Technomoral Politics of NGOs, Social Movements, and the State in India." *American Ethnologist* 43 (1): 76–90. https://doi.org/10.1111/amet.12264.

Bourdieu, Pierre. 1990. *The Logic of Practice*. Cambridge: Polity.

———. 1995. *Outline of a Theory of Practice.* Cambridge: Cambridge University Press. Originally published in 1972.

Bourne, Richard. 1978. *Assault on the Amazon.* London: Victor Gollancz.

Branford, Sue, and Oriel Glock. 1985. *The Last Frontier: Fighting over Land in the Amazon.* London: Zed Books.

Bratman, Michael. 1999. *Faces of Intention: Selected Essays on Intention and Agency.* Cambridge: Cambridge University Press.

Braudel, Fernand. 1990a. *The Mediterranean and the Mediterranean World in the Age of Philip II.* 2 vols. Vol. 1. New York: Harper and Row. Originally published in 1972.

———. 1990b. *The Mediterranean and the Mediterranean World in the Age of Philip II.* 2 vols. Vol. 2. New York: Harper and Row. Originally published in 1972.

Brenneis, Donald. 1994. "Discourse and Discipline at the National Research Council: A Bureaucratic Bildungsroman." *Cultural Anthropology* 9 (1): 23–36.

Brettell, Caroline B. 2002. "The Individual/Agent and Culture/Structure in the History of the Social Sciences." *Social Science History* 26 (3): 429–45.

Britan, Gerald M. 1981. *Bureaucracy and Innovation: An Ethnography of Policy Change.* Beverly Hills, CA: SAGE.

Broodbank, Cyprian. 2013. *The Making of the Middle Sea: A History of the Mediterranean from the Beginning to the Emergence of the Classical World.* London: Thames and Hudson.

Browder, John D., and Brian J. Godfrey. 1997. *Rainforest Cities: Urbanization, Development, and Globalization of the Brazilian Amazon.* New York: Columbia University Press.

Brown, Hannah, Adam Reed, and Thomas Yarrow. 2017. "Introduction: Towards an Ethnography of Meeting." *Journal of the Royal Anthropological Institute* 23 (S1): 10–26. https://doi.org/10.1111/1467-9655.12591.

Brown, Katrina, and Sergio Rosendo. 2000. "Environmentalists, Rubber Tappers, and Empowerment: The Politics and Economics of Extractive Reserves." *Development and Change* 31 (1): 201–27.

Bunker, Stephen G. 1988. *Underdeveloping the Amazon: Extraction, Unequal Exchange, and the Failure of the Modern State.* Chicago: University of Chicago Press.

Burdick, John. 1993. *Looking for God in Brazil: The Progressive Catholic Church in Urban Brazil's Religious Arena.* Berkeley: University of California Press.

Busby, Joshua William. 2007. "Bono Made Jesse Helms Cry: Jubilee 2000, Debt Relief, and Moral Action in International Politics." *International Studies Quarterly* 51 (2): 247–75.

Butler, Judith. 2015. *Notes toward a Performative Theory of Assembly.* Cambridge, MA: Harvard University Press.

Cabo, Isabel de. 2001. *La resistencia cultural bajo el franquismo: En torno a la revista "Destino" (1957–1961).* Barcelona: Áltera.

Capel, Horacio. 2005. *El modelo Barcelona: Un examen crítico.* Barcelona: Ediciones del Serbal.

Castellar-Gassol, Joan. 2000. *Barcelona, La història: Cròniques de 2000 anys de vida de la ciutat.* Barcelona: Edicions de 1984.

Castells, Manuel. 1989. *The Informational City: Information Technology, Economic Restructuring, and the Urban-Regional Process*. Oxford: Basil Blackwell.

———. 2008. "The New Public Sphere: Global Civil Society, Communication Networks, and Global Governance." *ANNALS of the American Academy of Political and Social Science* 616 (1): 78–93.

———. 2010. *The Rise of the Network Society*. Oxford: Wiley-Blackwell.

Castro, Eduardo Batalha Viveiros de. 1992. *From the Enemy's Point of View: Humanity and Divinity in an Amazonian Society*. Chicago: University of Chicago Press.

———. 1998. "Cosmological Deixis and Amerindian Perspectivism." *Journal of the Royal Anthropological Institute* 4 (3): 469–88.

———. 2004. "Exchanging Perspectives: The Transformation of Objects into Subjects in Amerindian Ontologies." *Common Knowledge* 10 (3): 463–84.

CATAC. 2013. "Qui som: Candidatura Autònoma de Treballadors i Treballadores de l'Administració de Catalunya." Accessed June 11. www.catac.cat.

CBD. 1992. "Convention on Biological Diversity." New York: United Nations.

CCCB. 2013. "Centre de Cultura Contemporània de Barcelona." Accessed February 16. www.cccb.org/en.

Chadwick, Simon, and Dave Arthur. 2008. "Més que un club (more than a club): The Commercial Development of FC Barcelona." In *International Cases in the Business of Sport*, edited by Simon Chadwick and Dave Arthur, 1–12. Amsterdam: Butterworth-Heinemann.

Chandler, David. 2014. "Democracy Unbound? Non-linear Politics and the Politicization of Everyday Life." *European Journal of Social Theory* 17 (1): 42–59.

Chapin, Marc. 2004. "A Challenge to Conservationists." *Worldwatch*. WorldWatch Institute.

Cheater, Angela. 1999. *The Anthropology of Power: Empowerment and Disempowerment in Changing Structures*. London: Routledge.

Childs, Sarah, and Mona Lena Krook. 2006. "Should Feminists Give Up on Critical Mass? A Contingent Yes." *Politics and Gender* 2 (4): 552–30.

———. 2009. "Analysing Women's Substantive Representation: From Critical Mass to Critical Actors." *Government and Opposition* 44 (2): 125–45.

Choudry, Aziz, and Dip Kapoor. 2013. *NGOization: Complicity, Contradictions, and Prospects*. London: Zed Books.

Ciemen. 2013. "Centre Internacional Escarré per las minories ètniques i les nacions." Accessed February 12. www.ciemen.cat/en.

Cirici, Alexandre, and A. Mercè Varela. 1975. *Més que un club: 75 anys del F. C. Barcelona*. Barcelona: Destino.

Cirtautas, Arista Maria. 1997. *The Polish Solidarity Movement: Revolution, Democracy, and Natural Rights*. London: Routledge.

Cleary, David. 1990. *Anatomy of the Amazon Gold Rush*. London: Macmillan and St. Antony's College.

———. 1993. "After the Frontier: Problems with Political Economy in the Modern Brazilian Amazon." *Journal of Latin American Studies* 25 (2): 331–49.

Comaroff, Jean, and John L. Comaroff. 1999. Introduction to *Civil Society and the*

Political Imagination in Africa: Critical Perspectives, edited by Jean Comaroff and John L. Comaroff. Chicago: University of Chicago Press.

Connolly, William E. 2011. *A World of Becoming*. Durham, NC: Duke University Press.

Conversi, Daniele. 1990. "Language or Race?: The Choice of Core Values in the Development of Catalan and Basque Nationalisms." *Ethnic and Racial Studies* 13 (1): 50–70. https://doi.org/10.1080/01419870.1990.9993661.

Corsín Jiménez, Alberto, and Adolfo Estalella. 2017. "Political Exhaustion and the Experiment of Street: Boyle Meets Hobbes in Occupy Madrid." *Journal of the Royal Anthropological Institute*. Special issue, *Meetings: Ethnographies of Organizational Process, Bureaucracy and Assembly* 23 (S1): 110–23.

Costa, Larissa, Viviane Junqueira, Cássio Martinho, and Jorge Fecuri. 2003. *Redes: Uma introdução às dinâmicas da conectividade e da auto-organização*. Brasília: WWF Brasil.

Crameri, Kathryn. 2000. *Language, the Novelist, and National Identity in Post-Franco Catalonia*. Oxford: Legenda.

Dagnino, Evelina. 2002. "Sociedade civil e espaços públicos no Brasil." São Paulo: Paz e Terra.

Davis, Wade. 1997. *One River: Science, Adventure and Hallucinogenics in the Amazon Basin*. New York: Simon and Schuster.

Davison, Peter. 2001. *Orwell in Spain: The Full Text of* Homage to Catalonia, *with Associated Articles, Reviews, and Letters from* The Complete Works of George Orwell. London: Penguin.

Dean, Mitchell. 2013. *The Signature of Power: Sovereignty, Governmentality, and Biopolitics*. Los Angeles: SAGE.

Dean, Warren. 1987. *Brazil and the Struggle for Rubber: A Study in Environmental History*. Cambridge: Cambridge University Press.

DeLanda, Manuel. 2006. *A New Philosophy of Society: Assemblage Theory and Social Complexity*. New York: Continuum.

———. 2010a. "Assemblages and Human History." In *Deleuze: History and Science*, 3–27. New York: Atropos Press.

———. 2010b. *Deleuze: History and Science*. New York: Atropos Press.

———. 2010c. "Metallic Assemblages." In *Deleuze: History and Science*, 67–80. New York: Atropos Press.

———. 2011. *A Thousand Years of Nonlinear History*. New York: Zone Books. Originally published in 1997.

Deleuze, Gilles, and Claire Parnet. 1987. *Dialogues*. Translated by Hugh Tomlinson and Barbara Habberjam. London: Athlone. Originally published in 1977.

Del Olmo, Carolina, and César Rendueles. 2004. "Las grietas de la ciudad capitalista: Entrevista con David Harvey." *Archipiélago: Cuadernos de Crítica de la Cultura* 62.

Delvaux, Bernard, and Eric Manez. 2008. "Towards a Sociology of the Knowledge-Policy Relation." In *Literature Review Integrative Report*. Knowledge and Policy. www.knowandpol.eu.

DeMars, William E. 2005. *NGOs and Transnational Networks: Wild Cards in World Politics*. London: Pluto Press.

Dezalay, Yves, and Bryant G. Garth. 1996. *Dealing in Virtue: International Commercial Arbitration and the Construction of a Transnational Legal Order*. Chicago: University of Chicago Press.

Díaz-Salazar, Rafael. 2002. Introduction to *Justicia global: Las alternativas de los movimientos del Foro de Porto Alegre*, edited by Rafael Díaz-Salazar. Barcelona: Icaria Editorial / Intermón Oxfam.

Douglas, Mary, and Aaron Wildavsky. 1982. *Risk and Culture: An Essay on the Selection of Technological and Environmental Dangers*. Berkeley: University of California Press.

Dryzek, John S. 1999. "Transnational Democracy." *Journal of Political Philosophy* 7 (1): 30–51.

Dumont, Louis. 1980. *Homo hierarchicus: The Caste System and Its Implications*. London: Weidenfeld and Nicholson. Originally published in 1966.

Durkheim, Émile. 2001. *The Elementary Forms of Religious Life*. Oxford: Oxford University Press.

Eagleton-Pierce, Matthew. 2001. "The Internet and the Seattle WTO Protests." *Peace Review* 13 (3): 331–37.

Edelman, Marc. 2001. "Social Movements: Changing Paradigms and Forms of Politics." *Annual Review of Anthropology* 30:285–317.

———. 2005. "When Networks Don't Work: The Rise and Fall and Rise of Civil Society Initiatives in Central America." In *Social Movements: An Anthropological Reader*, edited by June Nash. Oxford: Blackwell Publishing.

Eden, Michael J. 1990. *Ecology and Land Management in Amazonia*. London: Bellhaven Press.

Eder, Klaus. 2009. "The Making of a European Civil Society: 'Imagined,' 'Practised,' and 'Staged.'" *Policy and Society* 28 (1): 23–33.

Edles, Laura Desfor. 2010. *Symbol and Ritual in the New Spain*. Cambridge: Cambridge University Press.

Emirbayer, Mustafa, and Ann Mische. 1998. "What Is Agency?" *American Journal of Sociology* 103 (4): 962–1023.

Encarnación, Omar. 2003. *The Myth of Civil Society: Social Capital and Democratic Consolidation in Spain and Brazil*. New York: Palgrave Macmillan.

Enfield, Nick, and Paul Kockelman. 2017. *Distributed Agency*. New York: Oxford University Press.

Eriksen, Thomas Hylland. 1993a. *Ethnicity and Nationalism: Anthropological Perspectives*. London: Pluto Press.

———. 1993b. "In Which Sense Do Cultural Islands Exist?" *Social Anthropology* 1 (18): 133–47.

———. 2004. *What Is Anthropology?* London: Pluto Press.

Eriksen, Thomas Hylland, and Finn Sivert Nielsen. 2013. *A History of Anthropology*. 2nd ed. London: Pluto Press. Originally published in 2001.

Escobar, Arturo. 1992. "Culture, Economics, and Politics in Latin American Social

Movements Theory and Research." In *The Making of Social Movements in Latin America: Identity, Strategy, and Democracy*, edited by Arturo Escobar and Sonia E. Alvarez. Boulder, CO: Westview Press.

Farías Zurita, Víctor. 2009. *El mas i la vila a la Catalunya medieval: Els fonaments d'una societat senyorialitzada (segles XI–XIV)*. Valencia: Universitat de Valencia.

Fawcett, Paul, Matthew Flinders, Colin Hay, and Matthew Wood. 2017. *Anti-politics, Depoliticization, and Governance*. Oxford: Oxford University Press.

FCCD. 2013. "Fons Català de Cooperació al Desenvolupament." Accessed February 12. www.fonscatala.org.

Felman, Shoshana. 2003. *The Scandal of the Speaking Body: Don Juan with J. L. Austin, or Seduction in Two Languages*. Stanford, CA: Stanford University Press.

Ferguson, James. 1994. *The Anti-politics Machine: "Development," Depoliticization, and Bureaucratic Power in Lesotho*. Minneapolis: University of Minnesota Press.

Ferguson, James, and Akhil Gupta. 2002. "Spatializing States: Toward an Ethnography of Neoliberal Governmentality." *American Ethnologist* 29 (4): 981–1002.

Fisher, William F. 1997. "'Doing Good'? The Politics and Antipolitics of NGO Practices." *Annual Review of Anthropology* 26:439–64.

Fisher, William F., and Thomas Ponniah. 2003. *Another World Is Possible: Popular Alternatives in Globalization at the World Social Forum*. London: Zed Books.

Florensa i Soler, Núria. 1996. *El Consell de Cent: Barcelona a la guerra dels segadors*. Barcelona: Universitat Rovira i Virgili.

Fontova, Rosario. 2004. "Appelbaum defiende 'Voces,' la muestra de los 6 millones de euros." *El Periódico*, Barcelona.

Fortes, Meyer. 2004. *Kinship and the Social Order: The Legacy of Lewis Henry Morgan*. London: Routledge. Originally published in 1969.

ForumBCN. 2013. "Forum Barcelona 2004." Accessed February 12. www.barcelona2004.org.

Foucault, Michel. 1991. *Discipline and Punish: The Birth of the Prison*. London: Penguin. Originally published in 1975.

———. 2008. *The Birth of Biopolitics: Lectures at the Collège de France, 1978–1979*. London: Palgrave Macmillan.

Foweraker, Joe. 1974. "Political Conflict on the Frontier: A Case Study of the Land Problem in the West of Paraná, Brazil." PhD diss., University of Oxford.

———. 1981. *The Struggle for Land: A Political Economy of the Pioneer Frontier in Brazil from 1930 to the Present Day*. Cambridge: Cambridge University Press.

Friedman, Andrew L., and Samantha Miles. 2002. "Developing Stakeholder Theory." *Journal of Management Studies* 39 (1): 1–21.

FSMed. 2006. Technical Memory of the Mediterranean Social Forum. Barcelona: Technical Secretariat.

Furley, Peter A. 1994. *The Forest Frontier: Settlement and Change in Brazilian Roraima*. London: Routledge.

Garfield, Seth. 2004. "A Nationalist Environment: Indians, Nature, and the Construction of the Xingu National Park in Brazil." *Luso-Brazilian Review* 41 (1): 139–67.

Garland, Elizabeth. 1999. "Building Civil(ized) Society in the Kalahari and Beyond."

In *Civil Society and the Political Imagination in Africa: Critical Perspectives*, edited by Jean Comaroff and John L. Comaroff. Chicago: University of Chicago Press.

Gell, Alfred. 1998. *Art and Agency: An Anthropological Theory*. Oxford: Clarendon Press.

Gellner, David, and Mrigendra Bdr Karki. 2007. "The Sociology of Activism in Nepal: Some Preliminary Considerations." In *Social and Political Transformations in North India and Nepal: Social Dynamics in Northern South Asia*, edited by David Gellner and Katsuo Nawa, 361–97. Delhi: Manohar.

Gellner, Ernest. 1994. *Conditions of Liberty: Civil Society and Its Rivals*. London: Hamish Hamilton.

Gennep, Arnold van. 1960. *The Rites of Passage*. London: Routledge.

Giddens, Anthony. 1984. *The Constitution of Society: Outline of the Theory of Structuration*. Cambridge: Polity Press.

Gilbert, Margaret. 2006. *A Theory of Political Obligation: Membership, Commitment, and the Bonds of Society*. Oxford: Oxford University Press.

Ginsborg, Paul. 2008. *Democracy: Crisis and Renewal*. London: Profile Books.

Giugni, Marco, and Florence Passey. 1998. "Contentious Politics in Complex Societies." In *From Contention to Democracy*, edited by Marco Giugni, Doug McAdam, and Charles Tilly. Lanham, MD: Rowman and Littlefield.

Gledhill, John. 1994. *Power and Its Disguises: Anthropological Perspectives on Politics*. London: Pluto Press.

Gluckman, H. Max. 1954. *Rituals of Rebellion in South-East Africa: Frazer Lectures*. Manchester: Manchester University Press.

Goodman, David, and Anthony Hall. 1990. Introduction to *The Future of Amazonia: Destruction or Sustainable Development?*, edited by David Goodman and Anthony Hall. London: Macmillan.

Goodsell, Charles T. 2005. "The Bureau as Unit of Governance." In *The Values of Bureaucracy*, edited by Paul du Gay. Oxford: Oxford University Press.

Governance. 2007. "The Home of Good Governance." Accessed August 30, 2013. www.governance.co.uk.

Graeber, David. 2002. "The New Anarchists." *New Left Review* 13 (1): 61–73.

———. 2009. *Direct Action: An Ethnography*. Edinburgh: AK Press.

Gramsci, Antonio. 1971. *Selections from the Prison Notebooks*. Edited by Quintin Hoare and Geoffrey Smith. New York: International Publishers.

Grau, Ramón, and Marina López. 1988. *Exposición universal de Barcelona: Libro del centenario, 1888–1988*. Barcelona: Avenç.

Green, Judith, Maria Franquiz, and Carol Dixon. 1997. "The Myth of the Objective Transcript: Transcribing as a Situated Act." *TESOL Quarterly* 31 (1): 172–76.

Grosfoguel, Ramón. 2000. "Developmentalism, Modernity, and Dependency Theory in Latin America." *Nepantla: Views from South* 1 (2): 347–74.

Guibernau, Montserrat. 2012. *Catalan Nationalism: Francoism, Transition, and Democracy*. London: Routledge.

Habermas, Jürgen. 1981. *The Theory of Communicative Action*. Translated by Thomas McCarthy. 2 vols. Boston: Beacon Press.

Hall, Anthony. 1997. *Sustaining Amazonia: Grassroots Action for Productive Conservation*. Manchester: Manchester University Press.

———. 2000. "Amazonia at the Crossroads: The Challenge of Sustainable Development." London: Institute of Latin American Studies.

———. 2007. "Extractive Reserves: Building Natural Assets in the Brazilian Amazon." In *Reclaiming Nature: Environmental Justice and Ecological Restoration*, edited by James K. Boyce, Sunita Narain, and Elizabeth A. Stanton, 151–80. London: Anthem Press.

Hann, Chris. 1996. "Introduction: Political Society and Civil Anthropology." In *Civil Society: Challenging Western Models*, edited by Chris Hann and Elizabeth Dunn. European Association of Social Anthropologists. London: Routledge.

Hardt, Michael, and Antonio Negri. 2005. *Multitude: War and Democracy in the Age of Empire*. London: Harris Hamilton.

———. 2009. *Commonwealth*. Cambridge, MA: Harvard University Press.

Hargreaves, John. 2000. *Freedom for Catalonia? Catalan Nationalism, Spanish Identity, and the Barcelona Olympic Games*. Cambridge: Cambridge University Press.

Harris, Mark. 2010. *Rebellion on the Amazon: The Cabanagem, Race, and Popular Culture in the North of Brazil, 1798–1840*. Cambridge: Cambridge University Press.

Harrison, Rob, Terry Newholm, and Deirdre Shaw. 2005. *The Ethical Consumer*. Thousand Oaks, CA: SAGE.

Hartlieb, Susanne, and Bryn Jones. 2009. "Humanising Business through Ethical Labelling: Progress and Paradoxes in the UK." *Journal of Business Ethics* 88 (3): 583–600. https://doi.org/10.1007/s10551-009-0125-x.

Harvey, Penelope. 1996. *Hybrids of Modernity: Anthropology, the Nation State, and the Universal Exhibition*. London: Routledge.

Hawkins, Darren. 2002. "Human Rights Norms and Networks in Authoritarian Chile." In *Restructuring World Politics: Transnational Social Movements, Networks, and Norms*, edited by Sanjeev Khagram, James V. Riker, and Kathryn Sikkink. Minneapolis: University of Minnesota Press.

Hecht, Susanna. 1993. "The Logic of Livestock and Deforestation in the Amazon." *BioScience* 43 (10): 687–95.

Hecht, Susanna, and Alexander Cockburn. 1989. *The Fate of the Forest: Developers, Destroyers, and Defenders of the Amazon*. London: Verso.

Hemming, John. 1985a. "The Frontier after a Decade of Colonisation." In *Change in the Amazon Basin*. Manchester: Manchester Latin American Studies.

———. 1985b. "Man's Impact on Forests and Rivers." In *Change in the Amazon Basin*. Manchester: Manchester Latin American Studies.

———. 1987. *Amazon Frontier: The Defeat of the Brazilian Indians*. London: Macmillan.

———. 2005. "On the Death of Orlando Villas Boas and the Legacy of the Villas Boas Brothers." *Tipití: Journal of the Society for the Anthropology of Lowland South America* 3 (1): 91–101.

Hendricks, Frank. 2010. *Vital Democracy: A Theory of Democracy in Action*. Oxford: Oxford University Press.

Herzfeld, Michael. 1993. *The Social Production of Indifference: Exploring the Symbolic Roots of Western Bureaucracy.* Chicago: University of Chicago Press.

———. 1997. *Cultural Intimacy: Social Poetics in the Nation-State.* New York: Routledge.

Hirsch, Eric, and David Gellner. 2001. *Inside Organizations: Anthropologists at Work.* Oxford: Berg.

Hobart, Mark. 1993. *An Anthropological Critique of Development: The Growth of Ignorance.* London: Routledge.

Hochstetler, Kathryn, and Margaret E. Keck. 2007. *Greening Brazil: Environmental Activism in State and Society.* Durham, NC: Duke University Press.

Holloway, John. 2005. *Change the World without Taking Power.* London: Pluto Press.

Holston, James. 2008. *Insurgent Citizenship: Disjunctions of Democracy.* Princeton, NJ: Princeton University Press.

Hornborg, Alf. 2001. "Ecological Embeddedness and Personhood: Have We Always Been Capitalists?" In *Ecology and the Sacred: Engaging the Anthropology of Roy A. Rappaport,* edited by Ellen Messer and Michael Lambek. Ann Arbor: University of Michigan Press.

Horton, Dave. 2003. "Green Distinctions: The Performance of Identity among Environmental Activists." Supplement, *Sociological Review* 51 (S2): 63–77.

Hull, Matthew S. 2012. *Government of Paper: The Materiality of Bureaucracy in Urban Pakistan.* Berkeley: University of California Press.

Hulme, David, and Michael Edwards. 1997. *NGOs, States, and Donors: Too Close for Comfort?* New York: St. Martin's Press.

Huntington, Samuel P. 1993. "The Clash of Civilizations?" *Foreign Affairs* 72 (3): 22–49.

IEMed. 2013. "European Institute of the Mediterranean." Accessed February 12. www.iemed.org.

Ingold, Tim. 2000a. "Culture, Nature, Environment: Steps to an Ecology of Life." In *The Perception of the Environment: Essays on Livelihood, Dwelling, and Skill.* London: Routledge.

———. 2000b. *The Perception of the Environment: Essays on Livelihood, Dwelling, and Skill.* London: Routledge.

———. 2008. "Bindings against Boundaries: Entanglements of Life in an Open World." *Environment and Planning* 40 (8): 1796–810.

———. 2011. *Being Alive: Essays on Movement, Knowledge, and Description.* London: Routledge.

———. 2013. "Prospect." In *Biosocial Becomings: Integrating Social and Biological Anthropology,* edited by Tim Ingold and Gisli Palsson, 1–21. Cambridge: Cambridge University Press.

ISA. 2011. *Parque indígena do Xingu 50 anos, Almanaque Socioambiental.* São Paulo: Instituto Socioambiental.

Jackson, Joe. 2008. *The Thief at the End of the World: Rubber, Power, and the Seeds of Empire.* London: Duckworth Overlook.

James, Wendy. 2003. *The Ceremonial Animal: A New Portrait of Anthropology.* Oxford: Oxford University Press.

Jasanoff, Sheila. 2004. "The Idiom of Co-production." In *States of Knowledge: The Co-production of Science and the Social Order*, edited by Sheila Jasanoff, 1–12. London: Routledge.

Jasper, James M. 1997. *The Art of Moral Protest: Culture, Biography, and Creativity in Social Movements*. Chicago: University of Chicago Press.

———. 2011. "Emotions and Social Movements: Twenty Years of Theory and Research." *Annual Review of Sociology* 37:285–303.

Josselin, Daphne, and William Wallace. 2001. *Non-state Actors in World Politics*. Basingstoke: Palgrave.

Juris, Jeffrey S. 2008a. *Networking Futures: The Movements against Corporate Globalization*. Durham, NC: Duke University Press.

———. 2008b. "Performing Politics: Image, Embodiment, and Affective Solidarity during Anti-corporate Globalization Protests." *Ethnography* 9 (1): 61–97. https://doi.org/10.1177/1466138108088949.

Kamat, Sangeeta. 2004. "The Privatization of Public Interest: Theorizing NGO Discourse in a Neoliberal Era." *Review of International Political Economy* 11 (1): 155–76.

Kamau, Evanson C., and Gerd Winter. 2009. *Genetic Resources, Traditional Knowledge, and the Law: Solutions for Access and Benefit Sharing*. London: Earthscan.

Kamminga, Menno T. 2007. "What Makes an NGO 'Legitimate' in the Eyes of States?" In *NGO Involvement in International Governance and Policy*, edited by Anton Vedder et al., 175–95. Leiden: Martinus Nijhoff.

Keane, John. 2009. *The Life and Death of Democracy*. New York: Norton.

Keck, Margaret. 1995. "Parks, People, and Power: The Shifting Terrain of Environmentalism." *NACLA Report on the Americas* 28 (5): 36–43. https://doi.org/10.1080/10714839.1995.11725796.

Keck, Margaret, and Kathryn Sikkink. 1998a. *Activists beyond Borders: Advocacy Networks in International Politics*. Ithaca, NY: Cornell University Press.

———. 1998b. "Conclusions." In *Activists beyond Borders: Advocacy Networks in International Politics*, edited by Margaret Keck and Kathryn Sikkink, 199–218. Ithaca, NY: Cornell University Press.

———. 1998c. "Transnational Advocacy Networks in International Politics: Introduction." In *Activists beyond Borders: Advocacy Networks in International Politics*, edited by Margaret Keck and Kathryn Sikkink, 1–38. Ithaca, NY: Cornell University Press.

Kelly, Robert E. 2007. "From International Relations to Global Governance Theory: Conceptualizing NGOs after the Rio Breakthrough of 1992." *Journal of Civil Society* 3 (1): 81–99.

Kennelly, Jacqueline Joan. 2009. "Youth Cultures, Activism, and Agency: Revisiting Feminist Debates." *Gender and Education* 21 (3): 259–72.

Khagram, Sanjeev. 2002. "Restructuring the Global Politics of Development: The Case of India's Narmada Valley Dams." In *Restructuring World Politics*, edited by Sanjeev Khagram, James V. Riker, and Kathryn Sikkink. Minneapolis: University of Minnesota Press.

Khagram, Sanjeev, James V. Riker, and Kathryn Sikkink. 2002. *Restructuring World Politics: Transnational Social Movements, Networks, and Norms*. Minneapolis: University of Minnesota Press.

Kidder, Thalia G. 2002. "Networks in Transnational Labor Organizing." In *Restructuring World Politics*, edited by Sanjeev Khagram, James V. Riker, and Kathryn Sikkink. Minneapolis: University of Minnesota Press.

Kothari, Smitu. 2002. "Globalization, Global Alliances, and the Narmada Movement." In *Restructuring World Politics*, edited by Sanjeev Khagram, James V. Riker, and Kathryn Sikkink. Minneapolis: University of Minnesota Press.

Kousis, Maria, and Klaus Eder. 2001. "Introduction: EU Policy-Making, Local Action, and the Emergence of Institutions of Collective Action: A Theoretical Perspective on Southern Europe." *Environment and Policy* 29:3–21.

Krader, Lawrence. 1976. *Dialectic of Civil Society*. Assen: Van Gorcum.

Krøijer, Stine. 2015. *Figurations of the Future: Forms and Temporalities of Left Radical Politics in Northern Europe*. New York: Berghahn.

Kuper, Adam. 1999. *Culture: The Anthropologists' Account*. Cambridge, MA: Harvard University Press.

La Boétie, Étienne de, and Paul Bonnefon. 2007. *The Politics of Obedience and Étienne de la Boétie*. Montreal: Black Rose Books.

Lashaw, Amanda, Steven Sampson, and Christian N. Vannier, eds. 2016. *Cultures of Doing Good: Anthropologists and NGOs*. Tuscaloosa: University of Alabama Press.

Latour, Bruno. 1986. "The Powers of Association." In *Power, Action, and Belief: A New Sociology of Knowledge?*, edited by John Law, 264–80. London: Routledge and Kegan Paul.

———. 1993. *We Have Never Been Modern*. Cambridge, MA: Harvard University Press.

———. 2005. *Reassembling the Social: An Introduction to Actor-Network-Theory*. Oxford: Oxford University Press.

———. 2013. *An Inquiry into Modes of Existence: An Anthropology of the Moderns*. Cambridge, MA: Harvard University Press.

Layton, Robert. 2012. *An Introduction to Theory in Anthropology*. Cambridge: Cambridge University Press. Originally published in 1997.

Leve, Lauren, and Lamia Karim. 2001. "Introduction: Privatizing the State; Ethnography of Development, Transnational Capital, and NGOs." *Political and Legal Anthropological Review* 24 (1): 53–58.

Lévi-Strauss, Claude. 1978. *Myth and Meaning*. Routledge Classics. London: Routledge.

Lewis, David. 2008. "Crossing the Boundaries between 'Third Sector' and State: Life-Work Histories from the Philippines, Bangladesh, and the UK." *Third World Quarterly* 29 (1): 125–41. https://doi.org/10.1080/01436590701726582.

———. 2014. *Nongovernmental Organizations, Management and Development*. London: Routledge.

Lewis, David, and Paul Opoku-Mensah. 2006. "Moving Forward Research Agendas on International NGOs: Theory, Agency, and Context." *Journal of International Development* 18 (5): 665–75.

Lisansky, Judith. 1990. *Migrants to Amazonia: Spontaneous Colonization in the Brazilian Frontier*. Boulder, CO: Westview Press.

List, Christian, and Philip Pettit. 2011. *Group Agency: The Possibility, Design, and Status of Corporate Agents*. Oxford: Oxford University Press.

Little, Paul. 1995. "Ritual, Power, and Ethnography at the Rio Earth Summit." *Critique of Anthropology* 15 (3): 265–88.

Llobera, Josep R. 2004. *Foundations of National Identity: From Catalonia to Europe*. New York: Berghahn Books.

Macdonald, Terry. 2008. *Global Stakeholder Democracy: Power and Representation beyond Liberal States*. Oxford: Oxford University Press.

MacIntyre, Alasdair. 1998. *A Short History of Ethics*. London: Routledge Classics.

Maeckelbergh, Marianne. 2009. *The Will of the Many: How the Alterglobalisation Movement Is Changing the Face of Democracy*. London: Pluto Press.

Malinowski, Bronislaw. 1972. *Argonauts of the Western Pacific: An Account of Native Enterprise and Adventure in the Archipelagoes of Melanesian New Guinea*. London: Routledge and Kegan Paul. Originally published in 1922.

Marcus, George, and Erkan Saka. 2006. "Assemblage." *Theory, Culture, and Society* 23 (2–3): 101–6.

Markowitz, Lisa. 2001. "Finding the Field: Notes on the Ethnography of NGOs." *Human Organization* 60 (1): 40–46.

Marshall, Tim. 2004. *Transforming Barcelona*. London: Routledge.

Martí, Jordi Bonet, and Carolina del Olmo. 2004. "Barcelona: La reinvención de la ciudad portuaria en la nueva economía global." *Archipiélago: Cuadernos de Crítica de la Cultura* 62.

Martínez, Patricia, and Enrique González. 2013. "Toni Puig, la mano detrás del Modelo Barcelona." In *Magis*. Guadalajara: ITESO.

Masri, Safwan M. 2017. *Tunisia: An Arab Anomaly*. New York: Columbia University Press.

Massumi, Brian. 2002. *Parables of the Virtual: Movement, Affect, Sensation*. Durham, NC: Duke University Press.

Mattelart, Armand. 2000. *Networking the World: 1794–2000*. Minneapolis: University of Minnesota Press.

McCarthy, Helen, Paul Miller, and Paul Skidmore. 2004. *Network Logic: Who Governs in an Interconnected World?* London: Demos.

Meggers, Betty J. 1996. *Amazonia: Man and Culture in a Counterfeit Paradise*. Washington, DC: Smithsonian Institution Press.

Melucci, Alberto. 2003. *Challenging Codes: Collective Action in the Information Age*. Cambridge: Cambridge University Press. First edition 1996.

Mendes, Chico. 1991. *Fight for the Forest*. London: Latin American Bureau.

Mendizabal, Enrique. 2006. "Understanding Networks: The Functions of Research Policy Networks." In *Overseas Development Institute Working Paper Series*. London: ODI.

Mercator. 2013. "Mercator European Network of Language Diversity Centre." Accessed February 12. www.mercator-network.eu.

Mertz, Elizabeth, and Andria Timmer. 2010. "Introduction: Getting It Done; Ethnographic Perspectives on NGOs." *Political and Legal Anthropological Review* 33 (2): 171–77.

Mintz, Sidney. 1986. *Sweetness and Power: The Place of Sugar in Modern History.* New York: Penguin Books.

Monclús, Francisco. 2003. "El 'modelo Barcelona' ¿una fórmula original? De la 'reconstrucción' a los proyectos urbanos estratégicos (1997–2004)." *Perspectivas Urbanas* 18 (4): 399–421.

Monereo, Manuel, Miguel Riera, and Pep Valenzuela. 2002. *Hacia el partido de oposición: Foro Social Mundial/Porto Alegre 2002.* Barcelona: El Viejo Topo.

Montaner, Josep Maria. 2004a. "Argumentos de la Barcelona poliédrica." In *Barcelona 1992–2004*, edited by Guim Costa, 19–23. Barcelona: Gustavo Gili.

———. 2004b. "La evolución del modelo Barcelona (1979–2002)." In *Urbanismo en el siglo XXI: Bilbao, Madrid, Valencia, Barcelona; Una visión crítica*, edited by Jordi Borja and Zaida Muxí, 203–19. Barcelona: UPC ETSAB.

Moore, Henrietta. 2009. "Epistemology and Ethics: Perspectives from Africa." *Social Analysis* 53 (2): 207–18.

———. 2011. *Still Life: Hopes, Desires, and Satisfactions.* Cambridge: Polity Press.

Moran, Emilio F. 1990. "Private and Public Colonisation Schemes in Amazonia." In *The Future of Amazonia: Destruction or Sustainable Development?*, edited by David Goodman and Anthony Hall. London: Macmillan.

———. 1993. *Through Amazonian Eyes: The Human Ecology of Amazonian Populations.* Iowa City: University of Iowa Press.

Mosknes, Heidi, and Mia Melin. 2013. *Faith in Civil Society: Religious Actors as Drivers of Change.* Uppsala, Sweden: Uppsala Centre for Sustainable Development, Uppsala University.

Mosse, David. 2005. *Cultivating Development: An Ethnography of Aid Policy and Practice.* London: Pluto Press.

Mouffe, Chantal. 2005. *On the Political, Thinking in Action.* London: Routledge.

MSP-PPI. 2013. "Summary Sheet: Mediterranean Solar Plan Project Preparation Initiative." In *Neighbourhood Investment Facility*, edited by Union for the Mediterranean. Barcelona.

Mulgan, Geoff. 1997. *Connexity: How to Live in a Connected World.* London: Chatto and Windus.

———. 2006. *Good and Bad Power: The Ideals and Betrayals of Government.* London: Allen Lane.

Nash, Kate. 2008. "Global Citizenship as Showbusiness: The Cultural Politics of Make Poverty History." *Media, Culture, and Society* 30 (2): 167–81.

Navaro-Yashin, Yael. 2009. "Affective Spaces, Melancholic Objects: Ruination and the Production of Anthropological Knowledge." *Journal of the Royal Anthropological Institute* 15 (1): 1–18.

Navarro, Núria. 2004. "'Por la cultura, los hombres luchan y se matan': Entrevista con el sociólogo Dominique Walton." *El Periódico*, Barcelona.

Nelson, Paul J. 2002. "Agendas, Accountability, and Legitimacy among Transnational

Networks Lobbying the World Bank." In *Restructuring World Politics*, edited by Sanjeev Khagram, James V. Riker, and Kathryn Sikkink. Minneapolis: University of Minnesota Press.

Newman, Janet. 2005. "Bending Bureaucracy: Leadership and Multi-Level Governance." In *The Values of Bureaucracy*, edited by Paul du Gay. Oxford: Oxford University Press.

Nimtz, August. 2002. "Marx and Engels: The Prototypical Transnational Actors." In *Restructuring World Politics*, edited by Sanjeev Khagram, James V. Riker, and Kathryn Sikkink. Minneapolis: University of Minnesota Press.

Norbeck, Edward. 1963. "African Rituals of Conflict." *American Anthropologist* 65 (6): 1254–79.

Obrador, Pau, and Sean Carter. 2010. "Art, Politics, Memory: Tactical Tourism and the Route of Anarchism in Barcelona." *Cultural Geographies* 17 (4): 525–31.

O'Connor, Timothy. 1994. "Emergent Properties." *American Philosophical Quarterly* 31 (2): 91–104.

Olaveson, Tim. 2001. "Collective Effervescence and Communitas: Processual Models of Ritual and Society in Emile Durkheim and Victor Turner." *Dialectical Anthropology* 26 (2): 89–124. https://doi.org/10.1023/a:1020447706406.

Oliver, Pamela E., and Gerald Marwell. 1988. "The Paradox of Group Size in Collective Action: A Theory of the Critical Mass II." *American Sociological Review* 53 (1): 1–8.

Orlove, Benjamin S., and Stephen B. Brush. 1996. "Anthropology and the Conservation of Biodiversity." *Annual Review of Anthropology* 25 (1): 329–52. https://doi.org/10.1146/annurev.anthr0.25.1.329.

Orwell, George. 1938. *Homage to Catalonia*. London: Secker and Warburg.

———. 1945. *Animal Farm*. London: Secker and Warburg.

Ostrogorski, Mosei. 1974. *Democracy and the Party System in the United States*. New York: Arno Press.

Ostrom, Elinor. 1999. "Self-Governance and Forest Resources." CIFOR Occasional Paper no. 20. Center for International Forestry Research. Bogor, Indonesia: CIFOR.

Pace, Richard. 1998. *The Struggle for Amazon Town: Gurupá Revisited*. Boulder, CO: Lynne Rienner.

Pádua, José Augusto. 2012. "Environmentalism in Brazil: An Historical Perspective." In *A Companion to Global Environmental History*, edited by John Robert McNeill and Erin Stewart, 455–73. Oxford: Wiley-Blackwell.

———. 2013. "Nature and Territory in the Making of Brazil." *RCC Perspectives* 7:33–39.

Pagiola, Stefano, Joshua Bishop, and Natasha Landell-Mills. 2012. *Selling Forest Environmental Services*. Hoboken, NJ: Taylor and Francis.

Paley, Julia. 2008a. "Democracy: Anthropological Approaches." Santa Fe: School for Advanced Research Press.

———. 2008b. "Participatory Democracy, Transparency, and Good Governance in Ecuador: Why Have Social Movement Organizations at All?" In *Democracy:*

Anthropological Approaches, edited by Julia Paley. Santa Fe: School for Advanced Research Press.

Patterson, Mary. 2006. "Agency, Kinship, and History in North Ambrym." *Journal of the Royal Anthropological Institute* 12 (1): 211–17.

Pecqueur, Bernard, and Paulo Freire Vieira. 2015. "Territorial Resources and Sustainability: Analyzing Development in a 'Post-Fordist' Scenario." In *Transitions to Sustainability*, edited by François Mancebo and Ignacy Sachs, 141–58. Dordrecht: Springer.

Pellicer, Lluís. 2004. "Josep Ramoneda anuncia el fin del 'consenso urbano' en Barcelona." *El País*, Barcelona.

Petersen, John E., W. Michael Kemp, Rick Bartleson, Walter R. Boynton, Chung-Chi Chen, Jeffrey C. Cornwell, Robert H. Gardner, et al. 2003. "Multiscale Experiments in Coastal Ecology: Improving Realism and Advancing Theory." *BioScience* 53 (12): 1181–97.

Peterson, Kristin. 2001. "Benefit Sharing for All?: Bioprospecting NGOs, Intellectual Property Rights, New Governmentalities." *Political and Legal Anthropological Review* 24 (1): 78–91.

Pleyers, Geoffrey. 2010. *Alter-Globalization: Becoming Actors in the Global Age*. Cambridge: Polity.

Pollock, Graham. 2001. "Civil Society Theory and Euro-Nationalism." *Studies in Social and Political Thought* 4:31–56.

Posey, Darrell. 2004. *Indigenous Knowledge and Ethics: A Darrell Posey Reader*. Edited by Kristiana Plenderleith. New York: Routledge.

Power, Michael. 1997. *The Audit Society: Rituals of Verification*. Oxford: Oxford University Press.

Price, David. 1989. *Before the Bulldozer: The Nambiquara Indians and the World Bank*. Cabin John, MD: Seven Locks Press.

Puig Picart, Toni. 1992. "La ciutat de les associacions: Pensar, dirigir, gestionar i animar les asociacions dels ciutadans des del marketing de serveis." Edited by Institut Municipal d'Animació i Esplai. Barcelona: Ajuntament.

———. 1995. "Porque quiero mi asociación, la reinvento: O dónde introducir cambios organizativos para mejorar la calidad en las asociaciones de los voluntarios." Edited by Institut Municipal d'Animació i Esplai. Barcelona: Ajuntament.

———. 2003. *La comunicación municipal cómplice con los ciudadanos: Somos una marca de servicios pública con propuestas innovadoras y un estilo entusiasta*. Barcelona: Paidós.

Rabinow, Paul. 2007. "Midst Anthropology's Problems." In *Global Assemblages: Technology, Politics, and Ethics as Anthropological Problems*, edited by Aihwa Ong and Stephen J. Collier. Oxford: Blackwell.

Ramoneda, Josep. 2004. "War and Civilisation." In *At War*, edited by Josep Ramoneda. Barcelona: CCCB / Diputació de Barcelona.

Razsa, Maple, and Andrej Kurnik. 2012. "The Occupy Movement in Zizek's Hometown: Direct Democracy and a Politics of Becoming." *American Ethnologist* 39 (2): 238–58.

Redford, Kent H. 1991. "The Ecologically Noble Savage." *Cultural Survival Quarterly* 15 (1): 1–3.

Reed, Martin. 2005. "Beyond the Iron Cage? Bureaucracy and Democracy in the Knowledge Economy and Society." In *The Values of Bureaucracy*, edited by Paul du Gay. Oxford: Oxford University Press.

Reid, Elizabeth J. 2000. "Understanding the Word 'Advocacy': Context and Use." In *Structuring the Inquiry into Advocacy. Volume 1 of the Seminar Series Nonprofit Advocacy and the Policy Process*, edited by Elizabeth J. Reid. Washington, DC: Urban Institute.

Reventós, Manuel. 1987. *Els moviments socials a Barcelona en el segle XIX*. Barcelona: Crítica.

Rheingold, Howard. 2002. *Smart Mobs: The Next Social Revolution*. Cambridge: Perseus.

Ribeiro, Gustavo Lins. 2009. "Non-hegemonic Globalizations: Alter-Native Transnational Processes and Agents." *Anthropological Theory* 9 (3): 297–329.

Ricardo, Fany. 2004. *Terras indígenas e unidades de conservação da natureza: O desafio das sobreposições*. São Paulo: Instituto Socioambiental.

Richards, Audrey. 1971. "Introduction: The Nature of the Problem." In *Councils in Action*, edited by A. Richards and A. Kuper. Cambridge: Cambridge University Press.

Richards, Audrey, and Adam Kuper. 1971. *Councils in Action*. Cambridge: Cambridge University Press.

Riechmann, Jorge, and Francisco Fernández Buey. 1994. *Redes que dan libertad: Introducción a los nuevos movimientos sociales*. Barcelona: Paidós.

Riker, James V. 2002. "NGOs, Transnational Networks, International Donor Agencies, and the Prospects for Democratic Governance in Indonesia." In *Restructuring World Politics*, edited by Sanjeev Khagram, James V. Riker, and Kathryn Sikkink. Minneapolis: University of Minnesota Press.

Riles, Annelise. 2000. *The Network Inside Out*. Ann Arbor: University of Michigan Press.

———. 2006. *Documents: Artifacts of Modern Knowledge*. Ann Arbor: University of Michigan Press.

———. 2017. "Outputs: The Promises and Perils of Ethnographic Engagement after the Loss of Faith in Transnational Dialogue." *Journal of the Royal Anthropological Institute* 23 (S1): 182–97.

Ritchie, Mark. 2002. "A Practitioner's Perspective." In *Restructuring World Politics*, edited by Sanjeev Khagram, James V. Riker, and Kathryn Sikkink. Minneapolis: University of Minnesota Press.

Rival, Laura. 2006. "Amazonian Historical Ecologies." *Journal of the Royal Anthropological Institute* 12 (S1): S79–S94.

Rothbard, Murray N. 2007. "The Political Thought of Étienne de la Boétie." In *The Politics of Obedience and Étienne de la Boétie*, edited by Étienne de la Boétie and Paul Bonnefon, 1–32. Montreal: Black Rose Books.

Routledge, Paul. 2003. "Convergence Space: Process Geographies of Grassroots

Mobilization Networks." *Transactions of the Institute of British Geography*, n.s., 28:333–49.

Rutherford, Blair. 2004. "Desired Publics, Domestic Government, and Entangled Fears: On the Anthropology of Civil Society, Farm Workers, and White Farmers in Zimabwe." *Cultural Anthropology* 19 (1): 122–53.

Rutzen, Douglas. 2015. "Authoritarianism Goes Global (II): Civil Society under Assault." *Journal of Democracy* 26 (4): 28–39.

Sachs, Sybille, and Edwin Rühli. 2011. *Stakeholders Matter: A New Paradigm for Strategy in Society*. Cambridge: Cambridge University Press.

Sampson, Steven. 1996. "The Social Life of Projects: Importing Civil Society to Albania." In *Civil Society: Challenging Western Models*, edited by Chris Hann and Elizabeth Dunn. London: Routledge.

Santilli, Juliana. 2005. *Socioambientalismo e novos direitos: Proteção jurídica à diversidade biológica e cultural*. São Paulo: Fundação Peirópolis, IIEB, ISA.

Santos, Boaventura de Sousa. 2000. *A Crítica da Razão Indolente: Contra o desperdício da experiência*. São Paulo: Cortez.

———. 2005. *O Fórum Social Mundial: Manual de uso*. São Paulo: T. A. Queiroz.

———. 2006. *The Rise of the Global Left: The World Social Forum and Beyond*. London: Zed.

Santos, Roberto. 1980. *História econômica da Amazônia (1800–1920)*. São Paulo: T. A. Queiroz.

Savage, Mike. 2005. "The Popularity of Bureaucracy: Involvement in Voluntary Associations." In *The Values of Bureaucracy*, edited by Paul du Gay. Oxford: Oxford University Press.

Schmink, Marianne, and Charles H. Wood. 1992. *Contested Frontiers in Amazonia*. New York: Columbia University Press.

Schuller, Mark. 2009. "Gluing Globalization: NGOs as Intermediaries in Haiti." *Political and Legal Anthropological Review* 32 (1): 84–104.

———. 2012. *Killing with Kindness: Haiti, International Aid, and NGOs*. New Brunswick, NJ: Rutgers University Press.

Schwartzman, Helen B. 1987. "The Significance of Meetings in an American Mental Health Center." *American Ethnologist* 14 (2): 271–94. https://doi.org/10.1525/ae.1987.14.2.02a00060.

———. 1989. *The Meeting: Gatherings in Organizations and Communities*. Boston: Springer.

Scott, James. 1998. *Seeing like a State: How Certain Schemes to Improve the Human Condition Have Failed*. New Haven, CT: Yale University Press.

Seligman, Adam B., and Robert P. Weller. 2012. *Rethinking Pluralism: Ritual, Experience, and Ambiguity*. New York: Oxford University Press.

Seoane, José, and Emilio Taddei. 2001. *De Seattle a Porto Alegre*. Buenos Aires: CLACSO.

Shepard, Benjamin, L. M. Bogad, and Stephen Duncombe. 2008. "Performing vs. the Insurmountable: Theatrics, Activism, and Social Movements." *Liminalities: A Journal of Performance Studies* 4 (3): 1–30.

Shore, Chris, and Susan Wright. 1997. *Anthropology of Policy: Critical Perspectives on Governance and Power*. London: Routledge.

Sikkink, Kathryn. 1991. *Ideas and Institutions: Developmentalism in Brazil and Argentina*. Ithaca, NY: Cornell University Press.

Silva, Marina. 2005. "The Brazilian Protected Areas Program." *Conservation Biology* 19 (3): 608–11.

Skidmore, Thomas E. 1988. *The Politics of Military Rule in Brazil, 1964–1985*. New York: Oxford University Press.

Smismans, Stijn. 2003. "European Civil Society: Shaped by Discourses and Institutional Interests." *European Law Journal* 9 (4): 473–95.

Smith, Andrew. 2005. "Conceptualizing City Image Change: The 'Re-imagining' of Barcelona." *Tourism Geographies: An International Journal of Tourism Space, Place, and Environment* 7 (4): 398–423.

Soares-Filho, Britaldo Silveira, Daniel Curtis Nepstad, Lisa M. Curran, Gustavo Coutinho Cerqueira, Ricardo Alexandrino Garcia, Claudia Azevedo Ramos, Eliane Voll, Alice McDonald, Paul Lefebvre, and Peter Schlesinger. 2006. "Modelling Conservation in the Amazon Basin." *Nature* 440:520–23.

Souza, Marco. 1990. *O empate contra Chico Mendes*. São Paulo: Marco Zero.

Spencer, Jonathan. 2007. *Anthropology, Politics, and the State: Democracy and Violence in South Asia*. Cambridge: Cambridge University Press.

Spielberg, Steven. 1998. *Saving Private Ryan*. 169 min. USA: Paramount Pictures.

Srinivas, Nidhi. 2009. "Against NGOs?: A Critical Perspective on Nongovernmental Action." *Nonprofit and Voluntary Sector Quarterly* 38 (4): 614–26.

Strathern, Marilyn. 1990. *The Gender of the Gift: Problems with Women and Problems with Society in Melanesia*. Berkeley: University of California Press.

———. 1996. "Cutting the Network." *Journal of the Royal Anthropological Institute* 2 (3): 517–35.

Subirós, Pep. 1989. "Notas para una teoría de Barcelona." *Revista de Occidente* 97 (July): 99–110.

Sztompka, Piotr. 1990. "Agency and Revolution." *International Sociology* 5 (2): 129–44. https://doi.org/10.1177/026858090005002002.

———. 1991. *Society in Action: The Theory of Social Becoming*. Cambridge: Polity.

Takacs, David. 1996. *The Idea of Biodiversity: Philosophies of Paradise*. Baltimore: Johns Hopkins University Press.

Tarrow, Sidney. 1998. *Power in Movement: Social Movements and Contentious Politics*. 2nd ed. Cambridge: Cambridge University Press. Originally published in 1993.

Taylor, Rupert. 2002. "Interpreting Global Civil Society." *Voluntas: International Journal of Voluntary and Nonprofit Organizations* 13 (4): 339–47.

———. 2004. *Creating a Better World: Interpreting Global Civil Society*. Bloomfield: Kumarian Press.

Tembo, Fletcher. 2003. "The Multi-image Development NGO: An Agent of the New Imperialism?" *Development in Practice* 13 (5): 527–32.

Thorkelson, Eli. 2016. "The Infinite Rounds of the Stubborn: Reparative Futures at a French Political Protest." *Cultural Anthropology* 31 (4): 493–519.

Thrift, Nigel. 2008. *Non-representational Theory: Space, Politics, Affect.* London: Routledge.

Tilly, Charles. 1978. *From Mobilization to Revolution.* Reading, MA: Addison-Wesley.

Tilly, Charles, and Lesley J. Wood. 2009. *Social Movements: 1768–2008.* Boulder, CO: Paradigm.

Touraine, Alain. 1977. *The Self-Production of Society.* Chicago: University of Chicago Press.

———. 1981. *The Voice and the Eye: An Analysis of Social Movements.* Translated by Alan Duff. Cambridge: Cambridge University Press.

———. 1985. "An Introduction to the Study of Social Movements." *Social Research* 52 (4): 749–87.

Trallero, Manuel. 2004. *Barcelona 2004 como mentira!* Barcelona: Belacqua.

Tramullas, Gemma. 2004a. "La FAVB acusa al Fòrum de ser una 'maniobra especulativa.'" *El Periódico,* Barcelona.

———. 2004b. "'No' a la guerra y 'no' al Fòrum." *El Periódico,* Barcelona.

Traugot, Mark. 1994. *Repertoires and Cycles of Collective Action.* Durham, NC: Duke University Press.

Tsing, Anna Lowenhaupt. 2005. *Friction: An Ethnography of Global Connection.* Princeton, NJ: Princeton University Press.

Tvet, Terje. 1998. *Angels of Mercy or Development Diplomats? NGOs and Foreign Aid.* Oxford: James Currey, Africa World Press, and Red Sea Press.

UfM. 2013. "Union for the Mediterranean." Accessed February 12. www.ufm secretariat.org.

UN. 1948. "Universal Declaration of Human Rights." Accessed July 6, 2017. www. un.org.

UTE. 2004. *Barcelona marca registrada, un model per desarmar.* Edited by Unió Temporal d'Escribes. Barcelona: Virus.

Vadjunec, Jacqueline M., and Dianne Rocheleau. 2009. "Beyond Forest Cover: Land Use and Biodiversity in Rubber Trail Forests of the Chico Mendes Extractive Reserve." *Ecology and Society* 14 (2): 29 (online).

Van Dijk, Jan. 1999. *The Network Society: Social Aspects of New Media.* Thousand Oaks, CA: SAGE.

Varios. 2004. *La otra cara del Fòrum de les Cultures S.A.* Barcelona: Bellatera.

Vedder, Anton. 2007. *NGO Involvement in International Governance and Policy: Sources of Legitimacy.* Leiden: Martinus Nijhoff.

Vigil y Vázquez, Manuel. 1981. *Entre el franquismo y el catalanismo con Picasso en medio: Una crónica general del último y largo forcejeo entre la rígida sustentación del centralismo y el irreductible anhelo autonomista.* Barcelona: Plaza and Janés.

Villas Bôas, Orlando, and Claudio Villas Bôas. 1974. *Xingu: The Indians, Their Myths.* Translated by Susana H. Rudge. Edited by Kenneth Brechner. London: Souvenir Press.

Voulvouli, Aimilia. 2009. *From Environmentalism to Transenvironmentalism: The Ethnography of an Urban Protest in Modern Istanbul.* Bern: Peter Lang.

Wagley, Charles. 1958. *Amazon Town: A Study of Man in the Tropics.* New York: Macmillan.

Wallerstein, Immanuel. 1974. *The Modern World-System.* New York: Academic Press.

Warkentin, Craig. 2001. *Reshaping World Politics: NGOs, the Internet, and Global Civil Society.* Lanham, MD: Rowman and Littlefield.

Watts, Michael J. 1993. "Development I: Power, Knowledge, Discursive Practice." *Progress in Human Geography* 17 (2): 257–72. https://doi.org/10.1177/030913 259301700210.

WCED. 1987. *World Commission on Environment and Development Report: Our Common Future.* Oxford: Oxford University Press.

Wedel, Janine R. 1994. "U.S. Aid to Central and Eastern Europe, 1990–1994: An Analysis of Aid Models and Responses." In *East-Central European Economies in Transition*, edited by John Pearce Hardt and Richard F. Kaufman. Armonk, NY: M. E. Sharpe.

Weinstein, Barbara. 1983. *The Amazon Rubber Boom, 1850–1920.* Stanford, CA: Stanford University Press.

Whitaker, Chico. 2002. "Foro Social Mundial: Orígenes y objetivos." In *Porto Alegre: Otro mundo es posible*, edited by Manuel Monereo and Miguel Riera. Barcelona: El Viejo Topo.

———. 2005. *O desafio do Fórum Social Mundial: Um modo de ver.* São Paulo: Edições Loyola and Editora Fundação Perseu Abramo.

WHRC. 2013. "About Us." Woods Hole Research Center. Accessed June 24. whrc .org.

Wilkie, David S., Gilda A. Morelli, Josefien Demmer, Malcolm Starkey, Paul Telfer, and Matthew Steil. 2006. "Parks and People: Assessing the Human Welfare Effects of Establishing Protected Areas for Biodiversity Conservation." *Conservation Biology* 20 (1): 247–49.

Wolf, Eric R. 1999. *Envisioning Power: Ideologies of Dominance and Crisis.* Berkeley: University of California Press.

———. 2010. *Europe and the People without History.* Berkeley: University of California Press. Originally published in 1982.

World Bank. 2007. "What Is Our Approach to Governance?" Accessed July 26. www .worldbank.org/en/topic/governance/overview.

Wright, Susan. 1994. *Anthropology of Organizations.* New York: Routledge.

WSF. 2013. *Charter of Principles.* World Social Forum International Council 2002. Accessed June 17, 2013. www.forumsocialmundial.org.br.

Zigon, Jarrett. 2008. *Morality: An Anthropological Perspective.* Oxford: Berg.

Zubiri, Xavier. 2003. *Dynamic Structure of Reality.* Urbana: University of Illinois Press.

Index